AF581858

Small-Scale Energy Systems with Gas Turbines and Heat Pumps

Small-Scale Energy Systems with Gas Turbines and Heat Pumps

Editor

Satoru Okamoto

MDPI • Basel • Beijing • Wuhan • Barcelona • Belgrade • Manchester • Tokyo • Cluj • Tianjin

Editor
Satoru Okamoto
Shimane University
Japan

Editorial Office
MDPI
St. Alban-Anlage 66
4052 Basel, Switzerland

This is a reprint of articles from the Special Issue published online in the open access journal *Energies* (ISSN 1996-1073) (available at: https://www.mdpi.com/journal/energies/special_issues/small_scale_energy_systems).

For citation purposes, cite each article independently as indicated on the article page online and as indicated below:

LastName, A.A.; LastName, B.B.; LastName, C.C. Article Title. *Journal Name* **Year**, *Volume Number*, Page Range.

ISBN 978-3-0365-0072-0 (Hbk)
ISBN 978-3-0365-0073-7 (PDF)

Contents

About the Editor

Satoru Okamoto (Dr.) was a Professor in the Interdisciplinary Graduate School of Science and Engineering at Shimane University, Japan, from 1995 to 2017, and is currently Professor Emeritus. He obtained his M.Sc. in 1977 and Ph.D. in mechanical engineering in 1980 from Osaka University. Then, he worked as a researcher in the Research & Development Department at Ishikawajima-Harima Heavy Industries Co., Ltd. (IHI). At IHI, he worked as a researcher of aircraft engines and gas turbines and also designed compressors and fans of jet engines. He was an Associate Professor in the Department of Mechanical Engineering at Niihama National College of Technology from 1988 to 1994. Meanwhile, in 1994, he joined Professor Donald Rockwell's group at Lehigh University as a visiting research scientist. During this period, he explored an application of flow visualization and PIV in the field of flow-induced vibrations. Prof. Okamoto has served as Editorial Board Member for some leading energy journals and Chairperson of the 10th International Symposium on Advanced Science and Technology in Experimental Mechanics (10th ISEM '15-Matsue). In addition, he has published over 150 journal and conference publications, as well as 5 book chapters.

Preface to "Small-Scale Energy Systems with Gas Turbines and Heat Pumps"

This book results from a Special Issue published in *Energies*, entitled "Small-Scale Energy Systems with Gas Turbines and Heat Pumps". The purpose of this Special Issue is to provide information on innovation, research, development, and demonstration related to "Small-Scale Energy Systems with Gas Turbines and Heat Pumps". The main focuses of this Special Issue are conventional and non-conventional cooling, heating, and power technologies with gas turbines and heat pumps. Papers were solicited in areas including, but not limited to, the following: air conditioning, refrigeration, and heat pump systems; combined cycle, CHP, and CCHP with gas turbines; renewable energy for gas turbines and heat pumps; design and modeling/evaluation and optimization of small-scale energy systems with gas turbines and heat pumps.

Satoru Okamoto
Editor

Article

Harmonisation of Coolant Flow Pattern with Wake of Stator Vane to Improve Sealing Effectiveness Using a Wave-Shaped Rim Seal

Seungjin Lee, Daehan Kim and Joong Yull Park *

Department of Mechanical Engineering, Graduate School, Chung-Ang University, 84 Heukseok-ro, Dongjak-gu, Seoul 06974, Korea; leesj09@cau.ac.kr (S.L.); toto0825@cau.ac.kr (D.K.)
* Correspondence: jrpark@cau.ac.kr; Tel.: +82-2-820-5888

Received: 27 December 2018; Accepted: 10 March 2019; Published: 19 March 2019

Abstract: The rim seal of the gas turbine is intended to protect the material of the turbine disk from hot combustion gases. The study of the rim seal structure is important to minimise the coolant flow and maximise the sealing effect. In this paper, a wave-shaped rim seal for stator disks is proposed and its effect is confirmed by numerical analysis. To characterise the flow phenomena near the wave-shaped rim seal, a simplified model of the wave-shaped rim seal (Type 1 model), which excludes the rotor blade and stator vane, is analysed and compared with the conventional rim seal. Then, through analysis of the wave-shaped rim seal geometry (Type 2 model), which includes the rotor blade and stator vane, a reduction in egress and ingress flow was observed owing to the wave-shaped rim seal, and the sealing effectiveness on the stator disk of turbine was increased by up to 3.8%. Implementation of the wave-shape geometry in the radial seal is a novel choice for turbine designers to consider in future for better-performing and more-efficient turbines.

Keywords: wave-shaped rim seal; sealing effectiveness; radial seal; gas turbine; computational fluid dynamics

---

1. Introduction

To increase the thermal efficiency of the gas turbine, the inlet temperature of the gas needs to be increased [1]. However, hot gas reduces the turbine's lifespan, owing to thermal loading and fatigue failure of the turbine material. In preceding gas turbine studies, it was revealed that the cause of the ingress flow (hot mainstream gas) and egress flow through the wheelspace is the pressure difference near the interface between the mainstream gas path and the wheelspace [2] (Figure 1a). When passing through the mainstream gas path, the flow is affected by the wake of the stator vanes and rotor blades (Figure 1b(i)), causing non-axisymmetric variations in velocity and pressure (Figure 1b). In contrast, the pressure distribution in the wheelspace is relatively constant in the circumferential direction compared to that in the mainstream flow, resulting in a pressure difference between the two regions (Figure 1b(ii)). In modern gas turbines, the coolant flow is injected into the wheelspace to cool the turbine disk material and block the incoming hot gases (Figure 1a). However, there are two drawbacks to this approach. Since the coolant flow is bled from the compressor of the gas turbine system, the net efficiency of the gas turbine is reduced. In addition, the egress of the coolant flow through the gap between the stator and rotor disks interferes with the mainstream flow, and this deteriorates the aerodynamic performance. Therefore, the maximisation of cooling and sealing effects, while minimising coolant flow in gas turbines, has long been an important topic for engineers.

Basic studies have been conducted regarding the rotational flow in the wheelspace between the rotor and stator disks [3–5]; this was further developed into research on the ingress/egress of the flow through the wheelspace and how to minimise it. A mathematical method, called the orifice model, was devised to predict the minimum coolant flow required to prevent the ingress flow [6]. In addition, other methods, such as flow visualisation [7,8], gas concentration measurement [9,10] and computational fluid dynamics (CFD) [11,12], have been performed for studies on gas turbine disk design to minimise ingress/egress flows. When engineers design a gas turbine disk, a structure called a 'rim seal' is created around the disk to guide the flow between the wheelspace and the mainstream gas path (Figure 1a).

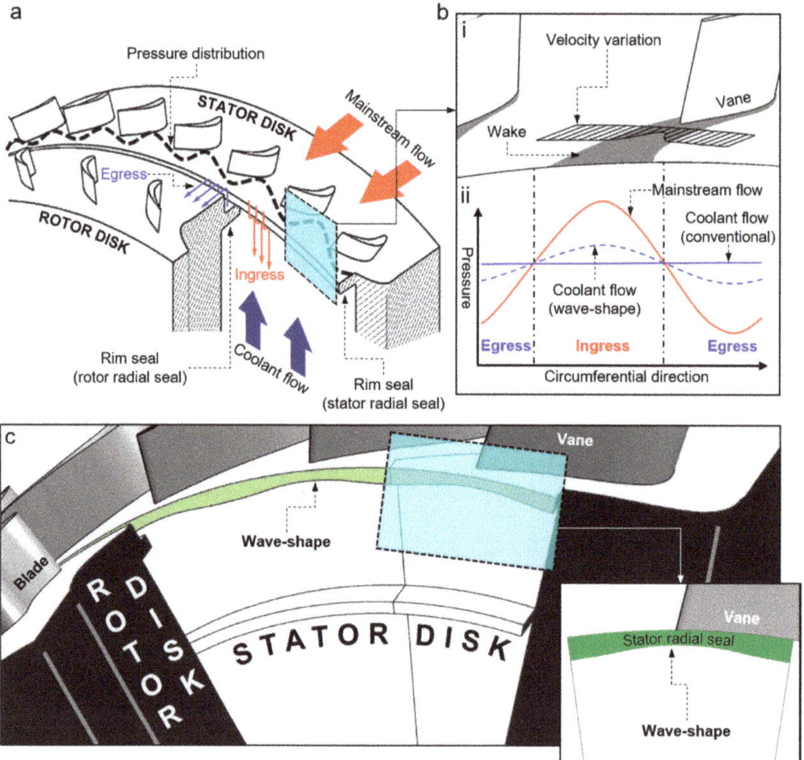

Figure 1. Schematics of the flow phenomenon around the rim seal and the concept of wave-shaped rim seal: (**a**) Flow phenomenon near the rim seal; (**b**) Pressure differences causing the egress and ingress flow. (**i**) Velocity profile of the mainstream flow; (**ii**) Pressure differences between the mainstream flow and coolant flow, and the expected effect of the wave-shaped rim seal; (**c**) Schematic view of the wave-shaped rim seal.

In the maximisation of cooling and sealing effects with a minimum coolant flow rate, there have been various concerns regarding the shape of the rim seal. Many studies are related to the deformation of the two-dimensional shape on the meridional section of the rim seal. Initial rim seal studies found that the sealing effect of the radial clearance seal was better than that of the axial clearance seal study [13]. A double radial clearance seal was then proposed and its performance verified [9,14]; this double radial clearance seal evolved into a study of the rim seal geometry called angel wings that divide the wheelspace into a trench cavity and a buffer cavity [15,16]. Additional studies were conducted to characterise the tendency of the flow around the rim seal according to the flow conditions (such as coolant flow rate, rotational Reynolds number, and rotational speed of the disk) [17] and the shape parameter of the rim seal [8,18]. Recently, a study was reported on the performance of a rim seal with an inner cavity and a rib structure added to the rotor side [19], and with two buffer cavities that attenuate the circumferential pressure asymmetries of the flow introduced from the mainstream gas path [20]. In contrast, a number of three-dimensional modified rim seals have been proposed. The protrusion attached to the rotor side of the wheelspace prevents ingestion of the mainstream flow by increasing the swirl ratio of the flow inside the wheelspace [21,22]. Similarly, a honeycomb-like rim seal geometry that promotes the sealing effect was also reported [23]. These studies on the three-dimensional shape of the rim seal are limited to the rotor-side rim seal, and contrarily, studies on the three-dimensional shape parameter of the stator rim seal are rare.

In this paper, a novel wave-shape geometry of the stator radial seal is proposed, which considers the three-dimensional shape parameter of the stator rim seal to improve the performance of the rim seal. The inner surface of the stator radial seal is formed to have a different radius along the circumference so that a wave-like cavity is created inside the stator radial seal (Figure 1c). Numerical analyses were performed for both the conventional and wave-shaped rim seal geometries, and the pressure distribution and velocity results are compared and discussed in detail. Sealing effectiveness and the ingress/egress flow structure are also confirmed by applying the flow condition of the coolant flow containing CO_2 gas in the simulation.

2. Materials and Methods

The analyses were conducted on two types of geometry (Figure 2a). Type 1 does not consider the stator vane and rotor blade to enable investigation of the sole effect of the rim seal geometries on flow dynamics, and the coolant flow channel is also omitted to save computational time. Type 2 considers the stator vane and rotor blade and includes the coolant flow channel to confirm the sealing effect of the wave-shaped rim seal geometry in the turbine.

2.1. Geometries

Most geometric dimensions used in this study are based on an experimental turbine at Arizona State University [24]. The outer radius of the disks is 195.7 mm, and the height of the mainstream gas path is 22.9 mm. The axial chord length of the rotor blade and stator vane is 31.8 mm and 48.3 mm, respectively, and the pitch of both the rotor blades and stator vanes is 59.2 mm. The tip clearance between the blade tip and the shroud is neglected. The profile of the stator vane and rotor blade consisted of profile points obtained from the midspan figure published in reference [24] using a data points extraction software (Engauge Digitizer, ver. 9.5, © 2014 Mark Mitchell). The axial gap between the rotor and stator disks is 16.5 mm. The axial clearance between the inner seal of the stator disk, located 153.2 mm from the axis of rotation, and the rotor disk is 2.5 mm (Figure 2b). The radial clearance seal has an axial overlap of 2 mm and a radial clearance of 2 mm. The length of the stator and rotor radial seals are 7.1 mm and 11.4 mm, respectively (Figure 2b). The coolant flow channel inlet located in the middle of the turbine has a radius of 19.1 mm. In the conventional rim seal, the radial clearance between the rotor and stator radial seals is constant at 2 mm (Figure 2c(i)). In contrast, in the wave-shaped rim seal, the stator radial seal is designed to have a wave-shape (green region in Figure 2c(ii)). At the ridge of the wave-shape, the gap is 3 mm, and the ridge is located at the same circumferential position as the trailing edge of the stator vane (Figure 2c(ii)). The gap at the troughs at both ends of the wave-shape is 1 mm (Figure 2c(ii)). A total of four geometries were used for the analysis (Figure 2d,e): Type 1 conventional rim seal (Type 1 C), Type 1 wave-shaped rim seal (Type 1 W), Type 2 conventional rim seal (Type 2 C), and Type 2 wave-shaped rim seal (Type 2 W). To save computation time, the geometry of a 360/22° sector model, including one stator vane and one rotor blade, was used for the analysis of a Type 2 turbine with 22 stator vanes and 22 rotor blades (Figure 2a). The 360/22° sector model was also applied to Type 1 (Figure 2a). To create the geometries, the BladeGen and DesignModeler programs of ANSYS 17.0 (Ansys, Inc., Canonsburg, PA, USA) were used.

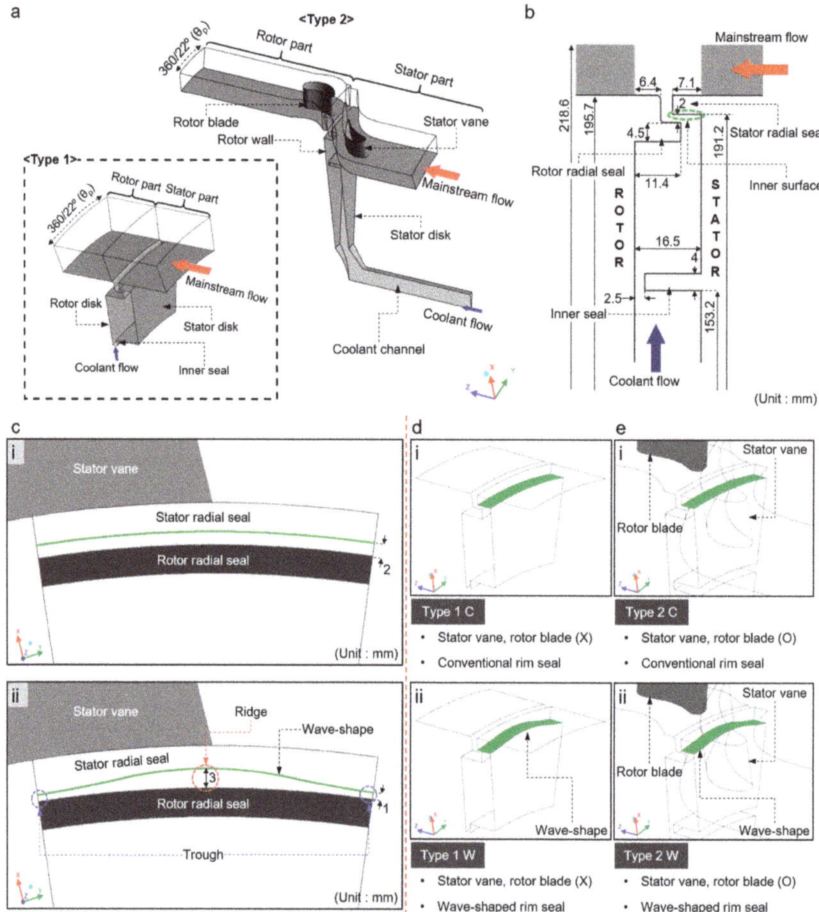

Figure 2. Schematics of geometries: (**a**) Computational domain of Type 1 and Type 2; (**b**) Dimensions of wheelspace; (**c**) front view of wheelspace; (**i**) the conventional rim seal and (**ii**) the wave-shaped rim seal; (**d**) inner surface of stator radial seal (green surface) of Type 1; (**i**) conventional rim seal and (**ii**) wave-shaped rim seal; (**e**) inner surface of stator radial seal (green surface) of Type 2; (**i**) conventional rim seal and (**ii**) wave-shaped rim seal.

2.2. Numerical Method and Boundary Conditions

Transient simulations were performed to predict the flow around the rim seal, which is dominated by time-dependent turbulent flow. A shear-stress transport (SST) k-ω model was used as the viscous model to accurately calculate the turbulent flow near the rim seal, considering the effect of the boundary layer formed on the wall of the narrow gap of the rim seal. The species transport model was applied to confirm the patterns of the egress and ingress flow, and the sealing effectiveness defined in terms of CO_2 gas concentration. The detailed mathematical formulas for SST k-ω model and species transport model are provided in 'Supplementary Materials'. ANSYS FLUENT 17.0, a commercial CFD tool, was used for the numerical analyses in this study. To generate the grid of the computational domain, ANSYS ICEM CFD 17.0 was used. The hexahedral grids were applied to all geometries. The same number of grids was used for the conventional rim seal and wave-shaped rim seal of each type (Figure 3a,b). In the wave-shaped rim seal, the radial seal grid number is the same as that of the conventional rim seal. However, as the radial gap changes owing to the wave shape of the wave-shaped rim seal, the grid becomes dense in the ridge region and sparse in the trough region. In the remaining parts of the wave-shaped rim seal, the same grid as that used for the conventional rim seal is used. In Type 1, the axial clearance, between the stator radial seal and the rotor wall, contains 53 cells and the radial clearance, between the stator radial seal and rotor radial seal, contains 48 cells (Figure 3c). There are 59 cells in the axial clearance between the rotor radial seal and the stator wall (Figure 3c). This is the same in Type 2. There are 49 and 84 cells in the circumferential direction in Type 1 and Type 2, respectively; this is because Type 2 contains stator -vanes and rotor blades. A total of 1,111,810 and 5,441,981 elements were used for Type 1 and Type 2, respectively. To predict the boundary layer formed on the stator vane and rotor blade, the widely used 'O-grid' scheme was applied to generate grids around these parts in Type 2 (Figure 3d). The values of y^+ are under five in the overall geometry; however, more refined grids ($y^+ < 1$) were used in the vicinity of the rim seal to predict the boundary layer accurately. A grid independence test was performed for the Type 1 W geometry. The appropriate grid density was determined based on the static pressure distribution in Line 1, located in the stator disk region inside the stator radial seal with the wave shape (Figure 4). The test was performed for the three different grid number cases (grid #1: 1.9×10^5, grid #2: 1.1×10^6, and grid #3: 3.3×10^6). The normalised angle, ζ (θ/θ_p), is defined as the angular position (θ) divided by the periodic angle ($\theta_p \approx 16.36°$) (Figure 4). Grids #2 and #3 exhibit almost the same P_{static} profile, whereas grid #1 has meaningfully different values. Therefore, the grid density and structure of grid #2 were applied to all models in this study (Figure 3).

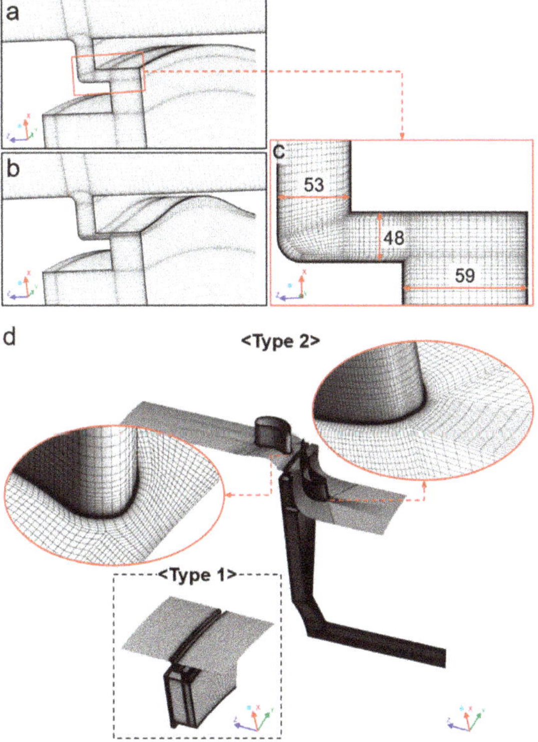

Figure 3. Computational grid. Grid for wheelspace of: (**a**) Type 1 C and (**b**) Type 1 W; (**c**) Enlarged view of grid for radial seal of Type 1 C; (**d**) Total view of grid for Type 1 and Type 2.

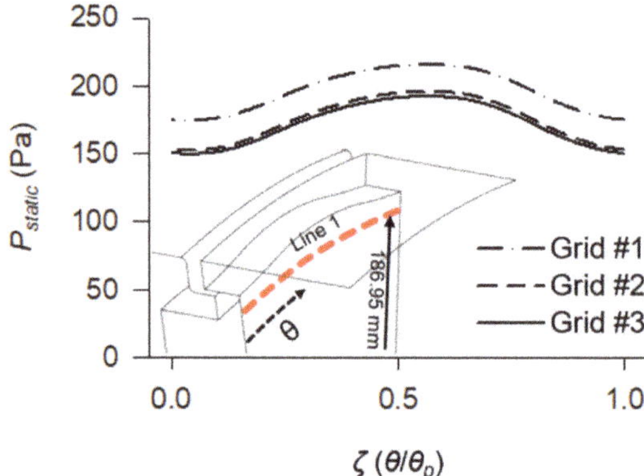

Figure 4. Grid independence test. Line 1 is located 186.95 mm from the axis of rotation on the stator disk region of Type 1 W.

Because the high-temperature environment inside the gas turbine is difficult to replicate in the laboratory, we also used CO_2 gas concentration measurements instead of high-temperature gas; this method has been employed often in both experimental [17,24] and numerical studies [23]. Our CFD model also used a mixture of air and CO_2 as the working fluid. The boundary conditions used in our models are based on the turbine experiments performed at Arizona State University [24], which are as follows. The mass flow rate of the mainstream flow was 0.11444 kg/s and contained only air, and the coolant flow at 0.24833×10^{-3} kg/s mass flow rate contained CO_2 gas at a mass fraction of 0.057. The outflow pressure condition was 97.5 kPa (absolute pressure). The rotating speed of rotor part was set to 2400 rpm. The periodic condition, with a $(360/22)^\circ$ ($\approx 16.36^\circ$) periodic angle of both the rotor part and stator part, was applied.

The time step was approximately 3.5×10^{-4} s for Type 1, and 3.5×10^{-5} s for Type 2, which is the time required for the rotor part to rotate by 5° (for Type 1) and 0.5° (for Type 2); note that the Type 2 geometry includes blades, but needs a more refined time step. The most important parameter with which to evaluate the rim seal performance in our study must be related to the mass fraction of CO_2, which stabilised after 2500 time steps for Type 1 models (Figure 5a) and 700 time steps for Type 2 models (Figure 5b). For the analysis of Type 1, the CO_2 data of the 5544th time step (when the stator and rotor parts were aligned in the initial position, shown in Figure 2a, Type 1) were used. In Type 2, the results of the 933rd–965th time steps were used; this time step period is the time required for the rotor to rotate a periodic angle of $360/22^\circ$ ($\approx 16.36^\circ$).

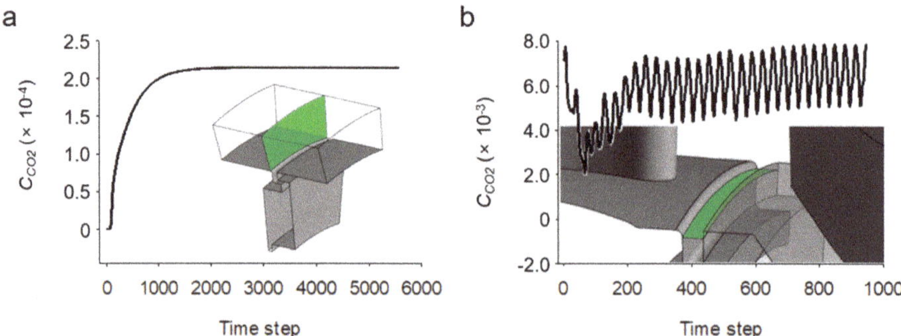

Figure 5. Monitoring of the area average mass fraction of CO_2 in green coloured plane to confirm flow stability for (**a**) Type 1 C and (**b**) Type 2 C.

3. Results and Discussion

Highly complex three-dimensional unsteady flow was found near the rim seal. The flow is affected by various flow dynamic issues, such as wake flow occurring near the stator vane [2], unsteady flow created by the rotor blade [23], and coolant flow, which were considered and achieved in Type 2 models (Figure 2). However, it is necessary to observe the sole influence of the wave-shaped rim seal geometry. Analysis of the Type 1 model, which excludes the stator vane and rotor blade (Figure 2), was preceded as a supportive model.

3.1. Type 1 (without Blade and Vane)

The most unique resultant effect of the wave-shaped rim seal is the uneven pressure distribution in the radial seal region (Figure 6a(ii)). A higher P_{static} forms around the ridge of the wave shape of the stator radial seal inner surface, even though there is no effect of the pressure field produced by the stator vane or rotor blade. It should be noted that the stator vane and rotor blade are not considered in Type 1 C, and thus there is an even pressure distribution (Figure 6a(i)). This tendency is confirmed by the comparison of the static pressure distributions at Line 2 (green point, Figure 6b), located on the inner surface of the stator radial seal of both the conventional and wave-shaped rim seals. In Line 2 of the conventional rim seal (Type 1 C), a uniform P_{static} formed at approximately 184.58 Pa (average P_{static}), in Line 2 of the wave-shaped rim seal (Type 1 W), a maximum P_{static} of 221.84 Pa formed at $\zeta = 0.57$ near the wave ridge and a minimum P_{static} of 143.60 Pa was confirmed $\zeta = 0.06$ in the vicinity of the wave trough (Figure 6b); this is a greater than $\pm$ 20% pressure difference to Type 1 C.

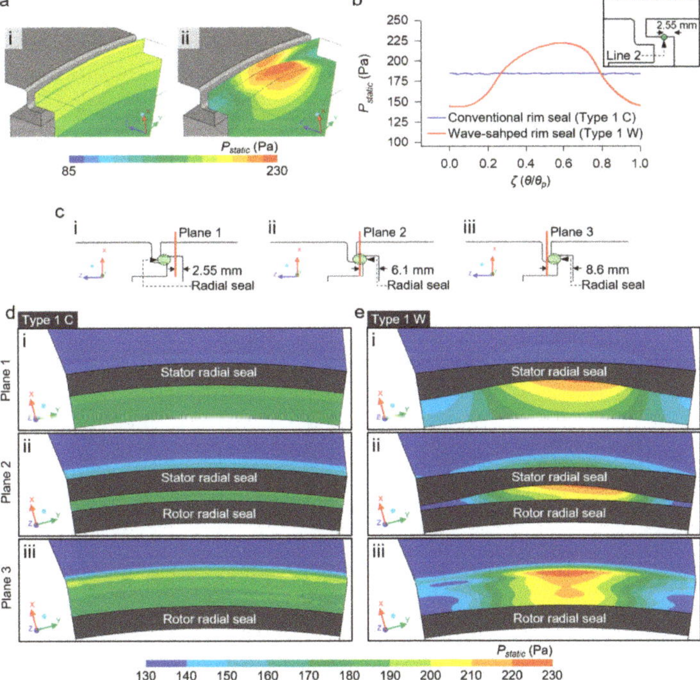

Figure 6. Static pressure distribution near the radius of the Type 1 wheelspace. (**a**) Static pressure contour of stator disk. (**i**) Type 1 C. (**ii**) Type 1 W. (**b**) Static pressure distribution along Line 2 (green point) on the inner surface of the stator radial seal. (**c**) Position of (**i**) Plane 1, (**ii**) Plane 2 and (**iii**) Plane 3. (**d**) Static pressure contour in (**i**) Plane 1, (**ii**) Plane 2, and (**iii**) Plane 3 of Type 1 C. (**e**) Static pressure contour in (**i**) Plane 1, (**ii**) Plane 2, and (**iii**) Plane 3 of Type 1 W.

For more detailed observation of the rim seal pressure profiles, Planes 1, 2, and 3 were set up at the cross-sections located at 2.55 mm, 6.1 mm, and 8.6 mm from the stator disk, respectively (Figure 6c), where the coolant and mainstream flows merge. In the wheelspace and mainstream path in Type 1 C, a uniform pressure along the circumferential direction was formed regardless of location (Figure 6d). Meanwhile, the uneven pressure distribution in the radial seal of Type 1 W (Figure 6a(ii)) propagates to the mainstream region. In Plane 1 of Type 1 W, high P_{static} was formed around the ridge of the wave shape (Figure 6e(i)) and in the mainstream region. This tendency is shown more clearly in Planes 2 and 3 (Figure 6e(ii,iii)); the peak pressure zone of 230 Pa permeates to the mainstream region (Figure 6e(iii)). In contrast, the counterpart images of Type 1 C (Figure 6d) show even pressure distributions with clearer circumferential boundaries between the rim seal and mainstream region. In Plane 3 of Type 1 C (Figure 6d(iii)) and Type 1 W (Figure 6e(iii)), fluctuating static pressure distributions were observed; this is caused by the rotation of the rotor disk. The pressure profiles on Plane 3 clearly show that the high pressure created in the radial seal invades the mainstream region and can change the mainstream pressure profile near the endwall.

Such dynamic change of the pressure profile found in Type 1 W (Figure 6e) is caused by the velocity distribution in the radial gap. Therefore, the velocity profiles were analysed in detail. In the radial gap, the circumferential velocity (V_{cir}) is dominant due to the rotation of the rotor. Planes 4, 5, 6, and 7 are the meridional sections at $\zeta = 0, 0.25, 0.5,$ and 0.75, respectively (Figure 7a). A high circumferential velocity (>36 m/s) was developed in the vicinity of the rotor disk wall for Type 1 C and Type 1 W (Figure 7b,c). In Type 1 C, the distribution of V_{cir} is same for Planes 4–7 (Figure 7b). This confirms that the flow pattern and the static pressure distribution (Figure 6d) of Type 1 C have the same uniform characteristics in the circumferential direction (θ-direction). On the contrary, in Type 1 W, different circumferential velocity distributions were formed for each plane (Planes 4–7) (Figure 7c). Different contours of V_{cir} were formed owing to the gradual change in the radial gap (distance between the stator radial seal and the rotor radial seal) determined by the wave-shaped rim seal geometry. Plane 4 (Figure 7c(i)) is a cross-section through a wave-shape trough and the higher circumferential velocity region (dotted line for >18 m/s) is larger (in terms of area portion) than those in Planes 5–7 (Figure 7c(ii–iv)). The circumferential velocity distribution is determined by the interplay between the rotor-side and stator-side walls. Unlike in Type 1 C (Figure 7b), the radial gap of Type 1 W (Figure 7c) changes continuously in the circumferential direction, and the interplay of the boundary layers of the two walls (stator and rotor walls) dynamically affect the V_{cir} distribution. As the radial gap widens, the region where V_{cir} is below 18 m/s (dashed line for <18 m/s) also widens (Figure 7c(i–iii)). Contrarily, as the radial gap narrows (in Plane 6), the area where $V_{cir} < 18$ m/s decreases, as shown in Figure 7c(iv). It is now important to observe how the velocity in the radial seal in Type 1 W modifies the static pressure profile. The area-averaged static pressure ($P_{ave,static}$) and area-averaged circumferential velocity ($V_{ave,cir}$) in the radial seal (Figure 8a(i), green) were calculated (Figure 8). The values of $P_{ave,static}$ and $V_{ave,cir}$ at $\zeta = 0, 0.25, 0.5, 0.75,$ and 1 (Figure 8a(ii)) show almost constant lines for both $P_{ave,static}$ ($\approx$ 176.88 Pa) and $V_{ave,cir}$ ($\approx$ 24.63 m/s) for Type 1 C (Figure 8b,c). However, in Type 1 W, the wave-shaped rim seal model, the peak $P_{ave,static}$ ($\approx$ 207.36 Pa) formed near the ridge of the wave shape ($\zeta = 0.5$) where the lowest $V_{ave,cir}$ ($\approx$ 22.56 m/s) was formed, whereas the lowest $P_{ave,static}$ ($\approx$ 119.39 Pa) and highest $V_{ave,cir}$ ($\approx$ 26.77 m/s) were formed in the trough ($\zeta = 0, 1$) (Figure 8b,c).

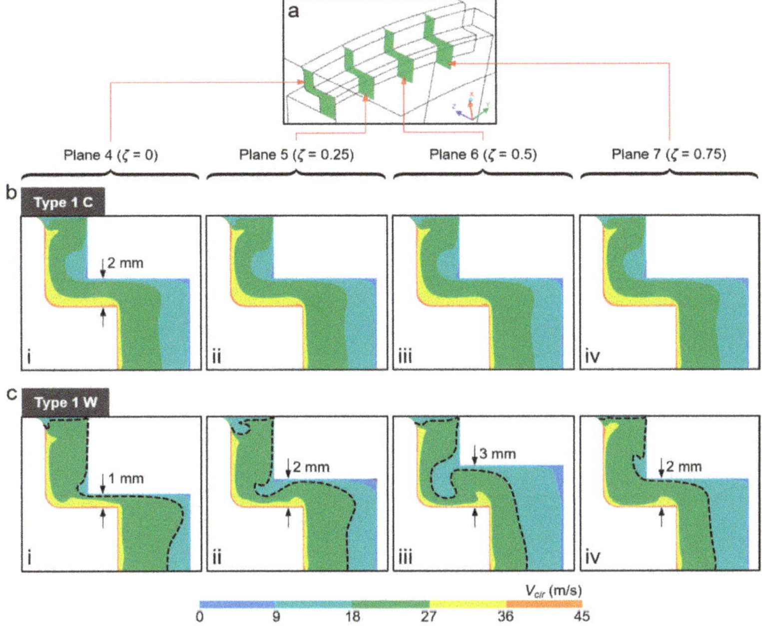

Figure 7. Circumferential velocity contour in wheelspace of Type 1. (**a**) Locations of Planes 4, 5, 6, and 7. (**b**) Circumferential velocity contour in wheelspace of Type 1 C in (**i**) Plane 4, (**ii**) Plane 5, (**iii**) Plane 6 and (**iv**) Plane 7. (**c**) Circumferential velocity contour in the wheelspace of Type 1 W in (i) Plane 4, (ii) Plane 5, (iii) Plane 6 and (**iv**) Plane 7.

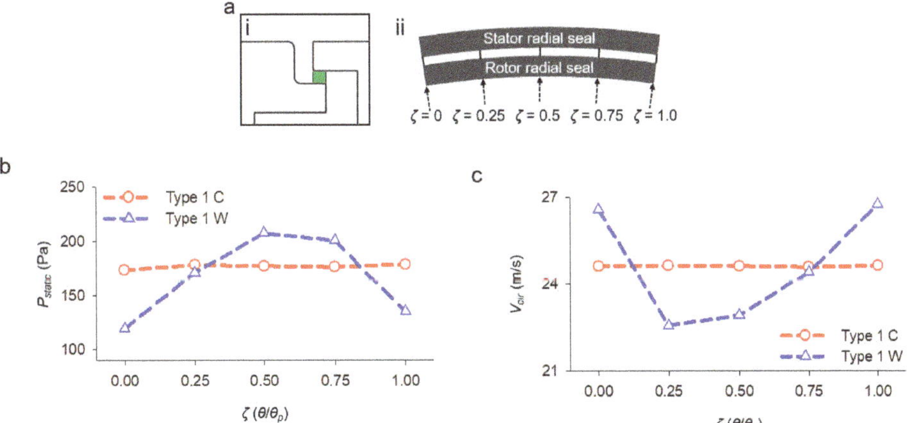

Figure 8. Numerical results of static pressure and circumferential velocity in the radial seal region. (**a**) Measuring plane position; (**b**) Static pressure distribution; (**c**) Circumferential velocity distribution.

In summary for Type 1 W, the rotation of the rotor disk in the radial seal is the main driver of the flow, and the greater the distance from the rotor disk, the lower the $V_{ave,cir}$. Therefore, near the ridge region with the wide radial gap of the wave-shaped rim seal, it forms a lower flow velocity than the velocity of the trough region, which causes the formation of high pressure near the ridge and low pressure near the trough. For Type 1 models, it is shown that the pressure distribution in the radial seal (Figure 6) can be controlled/designed. The rim seal prevents egress and ingress flows (Figure 1), and the pressure distribution downstream of the stator vane is strongly related to egress and ingress flows (Figure 1b). By controlling the pressure distribution in the radial seal through the wave-shaped rim seal geometry, it is expected that the egress and ingress flows can also be controlled.

3.2. Type 2 (with Blade and Vane)

Type 2 models include the stator vane and rotor blade geometries, and thus, exhibit fluid dynamic effects on egress and ingress flows. This section describes how the wave-shaped rim seal contributes to improving the sealing effect by reducing the egress and ingress flows. First, the Type 2 C model (conventional rim seal with blade and vane) was validated by comparing the simulation results with the referenced experimental result [24]. The absolute pressure (P_{abs}) distributions at the endwall 1 mm downstream of the stator vane (red point, Figure 9a(i)) and a point 186.95 mm from the axis of rotation (Figure 9a(i), blue point) were calculated. The calculated data at the red and blue points in Figure 9a(i) are plotted in ii and iii, respectively. In the referenced experimental result [24], asymmetric variations and uniform pressures were observed in the mainstream path and wheelspace, respectively (Figure 9a(ii)). Since these two different pressure distribution patterns significantly influence the development of ingress and egress flow near the rim seal [2], these two pressure patterns were simulated and analyzed in our CFD models (Figure 9a(iii)). The referenced experimental results show that the peak of the pressure distribution in the mainstream path was measured at $\zeta = 0.53$ near the stator vane trailing edge, and the pressure intersection point of the two pressure patterns was observed at $0.33 < \zeta < 0.4$ and $0.66 < \zeta < 0.73$ (Figure 9a(ii)). In the CFD model, the peak pressure in the mainstream path was located at $\zeta = 0.56$, and the points at which the two pressure patterns intersected were observed at $\zeta = 0.33$ and $\zeta = 0.68$, which are similar to the experimental results. However, there is a discrepancy in the pressure values predicted from the CFD model and the referenced experimental data. This is due to the inevitable inaccuracies that occur during the extraction of the stator vane and rotor blade geometry from reference [24] as described in Materials and Methods. Nevertheless, the observed pressure distribution trends of the CFD model are consistent with the experimental results, which supports the results and claims of this study.

Another important observation factor, the sealing effectiveness (ε_c) of CO_2 gas, is determined by the following equation [23]:

$$\varepsilon_c = (C_l - C_h)/(C_c - C_h) \tag{1}$$

where ε_c is the sealing effectiveness, C_l the local mass fraction of CO_2, C_h the mass fraction of CO_2 at the mainstream flow inlet, C_c the mass fraction of CO_2 at the coolant flow inlet. According to the radial position (r/R; r is the radial distance from the axis of the turbine and R is the disk radius) of the stator disk, both the experimental and CFD results show a similar tendency that ε_c decreases as the coolant passes through the inner seal ($r/R = 0.78$) and approaches the mainstream path (Figure 2b). Pressure distribution and sealing effectiveness comparisons exhibit a similar tendency in the CFD and experimental results.

The pressure distribution near the radial seal in the Type 2 models is considerably different from that in the Type 1 models because the mainstream flow passing the rotor blade and stator vane creates a disturbed pressure profile. Pressure patterns in the wheelspace are affected by the pressure distribution near the leading edge of the rotating rotor blade; this is observed at an interval of 4° rotation (= 8 time steps) of the rotor part (Figure 10a–c). During this process, the wave-shape geometry plays an important role in determining the pressure contours in the wheelspace. In detail, in the region of the Type 2 C radial seal, at Planes 1–3 (defined as in Figure 6c), the migration of pressure distribution was observed through the static pressure contour (Figure 10a–c(i)). The low pressure region represented by a white star migrated in the direction of rotation, and the next low pressure region (white point in Figure 10c(i)) was revealed after the existing low pressure (white star in Figure 10c(i)) had almost passed. A high-pressure region was formed between the two low pressure regions represented by the white circle and the star (Figure 10c(i)). Further, it is observed that the pressure profile shifted in Type 2 W, and near the ridge of radial seal, alternating high pressure and low pressure formed with a period corresponding to the rotation of the rotor blade (high pressure–Figure 10a(ii); low pressure–Figure 10b(ii); and high pressure–Figure 10c(ii) were found at the ridge). When the high pressure passed near the radial seal ridge (Figure 10a), in Plane 3, the area represented by a dashed line (> -1.4×10^{-3} Pa) for Type 2 W (Figure 10a(ii)) was formed larger (15.4%) than that in Type 2 C (Figure 10a(i)); the area for Type 2 W and Type 2 C was 0.15×10^{-3} and 0.13×10^{-3} m^2, respectively. When the low pressure passed near the radial seal ridge (Figure 10b), in Plane 3, the area represented by a dashed line (< -2.1×10^{-3} Pa) for Type 2 C was 4.49×10^{-5} m^2 (Figure 10b(i)), while the whole area (of Plane 3) was greater than -2.1×10^{-3} Pa for Type 2 W (Figure 10b(ii)). In Figure 10c, the high pressure passed near the ridge of wave shape again; the area marked with a dashed line (> -1.4×10^{-3} Pa) for Type 2 W was 4.55×10^{-5} m^2 (Figure 10c(ii)), while the whole area (of Plane 3) was below -1.4×10^{-3} Pa for Type 2 C (Figure 10c(i)). In summary, it is thus confirmed that the mainstream interplay changes the overall pressure profile near the rim seal; however, the increase in pressure near the ridge of the wave-shape geometry is constantly observable in both Type 1 W and Type 2 W. Therefore, the wave-shaped rim seal geometry contributes to the pressure profile by enhancing the performance of the rim seal.

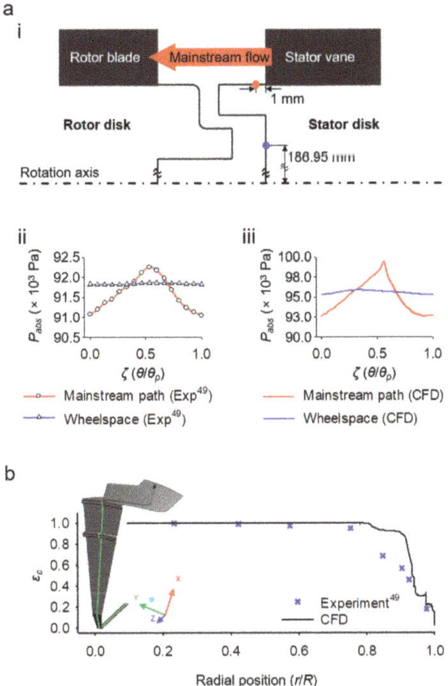

Figure 9. (**a**) Absolute pressure distribution. (**i**) Pressure measurement point of mainstream path (red point) and wheelspace (blue point). (**ii**) Absolute pressure distribution of experiment [24]. (**iii**) Absolute pressure distribution of CFD. (**b**) Sealing effectiveness along the mid-line (green line) of the stator disk.

The new pressure distribution profile created by the wave-shaped rim seal has the beneficial effect of decreasing the egress and ingress flow. Egress and ingress flow occur at the interface between the mainstream region and wheelspace owing to the pressure difference. For Type 2 W, the modified pressure field in the radial seal weakens the strength of the egress and ingress flow. The radial velocity component of egress and ingress flows, V_{eg} and V_{in}, in the interface between the mainstream path and wheelspace, respectively, were measured, while the rotor blade rotated by 16° (933rd–965th time steps in the simulation), which is equivalent to the approximate periodic angle (360/22°) (Figure 10d).

It should be noted that the maximum values of V_{eg} (; $V_{max,eg}$) and V_{in} (; $V_{max,in}$) change in each time step owing to the location change of the rotor (Figure 10d). Therefore, their averages over the 933rd–965th time steps were used; the detailed reason for using this range of time steps is addressed in the Methods section. The average of $V_{max,eg}$ for the 933rd–965th time steps was 38.13 m/s in Type 2 W, which is 1.12% lower than the 38.56 m/s average of $V_{max,eg}$ of Type 2 C. The average of $V_{max,in}$ for the same time steps was 32.41 m/s in Type 2 W, which is 2.67% lower than the 33.30 m/s average of $V_{max,in}$ of Type 2 C. This result means that the egress and ingress flow in Type 2 W is weaker than in Type 2 C. The stability of egress and ingress flow is expressed as the standard deviation of the velocity at each time step. The standard deviation of $V_{max,eg}$ was 9.77 in Type 2 C, and 7.21 in Type 2 W. The standard deviation of $V_{max,in}$ was 4.81 in Type 2 C, and 4.31 in Type 2 W. Lower standard deviations in Type 2 W imply that the egress and ingress flows are more stable and steady, and that the fluctuation of radial flow in this region is suppressed in Type 2 W. For a representative sample, the comparison of radial velocity contours in the interface is plotted in Figure 10e, which shows the reduction in egress and ingress flow due to the wave-shaped rim seal. The wave-shaped rim seal (Type 2 W) is thus useful to enhance the aerodynamic performance by reducing the egress and ingress flows and unwanted interaction between the mainstream and coolant flows.

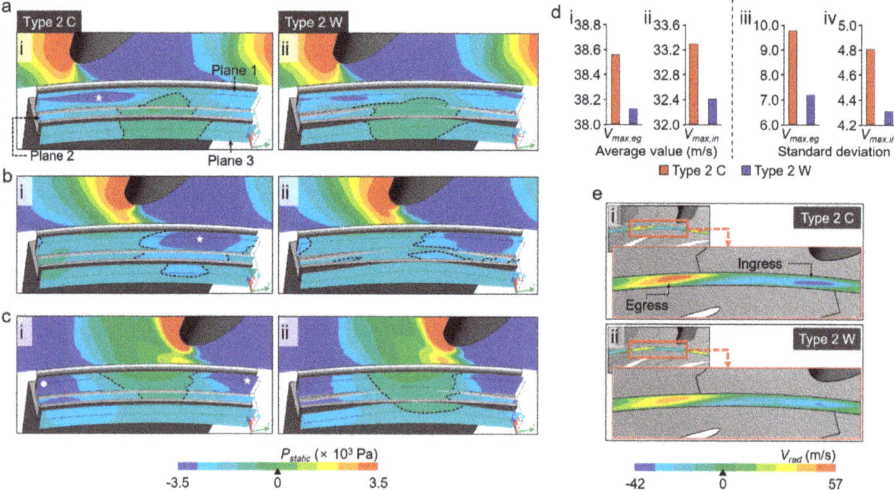

Figure 10. (**a–c**) Static pressure contours near the radial seal according to the rotation of the rotor blade. (**i**) Type 2 C. (**ii**) Type 2 W. (**d**) Reduction in egress and ingress flow due to the wave-shaped rim seal. (**i**) Average values of $V_{max,eg}$. (**ii**) Average values of $V_{max,in}$. (**iii**) Standard deviations of $V_{max,eg}$. (**iv**) Standard deviations of $V_{max,in}$. (**e**) Radial velocity contour at the interface between the mainstream path and wheelspace.

The degree of protection of the disk by the coolant flow was measured by comparing the mass fractions of CO_2 (C_{CO2}) on the surfaces of the rotor and stator disks for both the Type 2 C and W models (Figure 11a–d) according to the position of the rotor blade at an interval of 4° rotation of the rotor part (i–iv in Figure 11a–d). The C_{CO2} contour of the rotor disk is described in Region A (near the mainstream path) and Region B (near the rotor radial seal) (marked in Figure 11a(i)). In Region A, the cyan region ($0.009 < C_{CO2} < 0.018$ in Figure 11a,b(i–iv)) represents the egress flow path on the rotor disk side. In both Type 2 C and Type 2 W (Figure 11a,b, respectively), the coolant flow joins the mainstream path through the space between the rotor blades, because the high pressure near the leading edge of rotor blade (as shown in Figure 10) prevents the coolant flow from being directed near the leading edge of the rotor blade. Through comparison of the cyan areas between Type 2 C and W in Region A, it was observed that a higher C_{CO2} distribution was generated in Type 2 C (Figure 11a) than in Type 2 W (Figure 11b); this implies that a stronger egress flow is formed near the rotor disk of Type 2 C. In Region B, the red area (C_{CO2} > 0.045) is higher for Type 2 W (Figure 11b) than Type 2 C (Figure 11a). Through this comparison of the C_{CO2} contour patterns of Region B, where the rotor radial seal is located, for Type 2 C and Type 2 W, it was confirmed that the wave-shaped radial seal of Type 2 W improves the degree of protection of the rotor disk from the mainstream flow. This sealing effect improvement also appeared clearly on the surface of the stator disk (Figure 11c,d). In both Region A and Region B on the stator disk (marked in Figure 11c(i)), a higher C_{CO2} distribution was formed in Type 2 W than in Type 2 C. This also implies that the surface of the Type 2 W stator disk is better protected by the coolant flow than that of Type 2 C (Figure 11c,d). On the inner surface of the stator radial seal (green surface in Figure 2) of Type 2 C, the green region ($0.018 < C_{CO2} < 0.027$, Figure 11c) migrated depending on the location of the rotor blade (Figure 11c(i–iv)). However, in Type 2 W, this green region ($0.018 < C_{CO2} < 0.027$, Figure 11d) formed near the ridge of the wave-shaped radial seal and maintained its position despite the rotation of the rotor blade (Figure 11d(i–iv)). The change in CO_2 concentration distribution in Type 2 W is suppressed compared to that in Type 2 C, in accordance with the above-mentioned results (Figure 10d), which show that the implementation of the wave rim seal has an effect on suppressing the fluctuation of pressure and velocity.

Finally, the sealing effectiveness was quantitatively analysed. In Planes 8 and 9 (marked in Figure 12a(i)), the C_{CO2} was higher in Type 2 W (Figure 12a(ii)) than in Type 2 C (Figure 12a(i)). The sealing effectiveness in Planes 8 and 9 was calculated by Equation (1) using the average value of C_{CO2} calculated as the rotor blade rotates by 16° (933rd–965th time steps in the simulation), which is equivalent to an approximate periodic angle of (360/22°) (Figure 12b). Similar to the tendency observed in the comparison of C_{CO2} contours between Type 2 C and Type 2 W (in Figure 12a), in both Planes 8 and 9, the sealing effectiveness in Type 2 W was found to be higher than that in Type 2 C (Figure 12b). In Plane 8, the sealing effectiveness was 0.114 in Type 2 C and 0.119 in Type 2 W; the sealing effectiveness in Type 2 W was improved by 3.8% compared to Type 2 C (Figure 12b(i)). In Plane 9, the sealing effectiveness was 0.549 in Type 2 C and 0.560 in Type 2 W; the sealing effectiveness in Type 2 W was improved by 2.09% compared to Type 2 C (Figure 12b(ii)). An additional analysis was performed to quantify the coolant flow effectiveness in the rim seal space; the lines of R1–9 (green lines on the rotor disk in Figure 12c) and S1–9 (cyan lines on the stator disk in Figure 12c) were used to calculate ε_c on the walls of the rotor and stator disks. The average C_{CO2} values of the 933rd–965th time steps were used again for this analysis. On the rotor disk, in both Type 2 C and Type 2 W, it was observed that the sealing effectiveness was increased by the radial seal, and this tendency also appeared on the stator disk (Figure 12d,e). However, in R4–9 on the rotor disk, the sealing effectiveness of Type 2 W was higher than that of Type 2 C, and the difference in sealing effectiveness between Type 2 C and Type 2 W increased gradually with distance from the mainstream path (Figure 12d). R9 shows the biggest difference, where the sealing effectiveness in Type 2 W improved by 5.7% compared to Type 2 C. On the stator disk, the improvement in sealing effectiveness in Type 2 W was observed in S1–6, near the mainstream path (Figure 12e); this means that the protection of the stator disk by the wave-shaped radial seal in Type 2 W is improved in the region near the mainstream path. S4 shows the biggest difference between Type 2 C and Type 2 W, and the sealing effectiveness in Type 2 W improved by 17% compared to Type 2 C.

In this study, the analysis was performed on two tracks. The Type 1 model, which is the ideal model (no stator and rotor included), useful for understanding the fluid dynamic characteristics of the wave rim seal, was first analyzed to clearly quantify the effect of the wave rim seal. Then a more realistic model, the Type 2 model, with stator vanes and rotor blades included was performed to evaluate the role of the wave rim seal in highly complex unsteady flow caused by the interactions with the rotating stator and rotor blades.

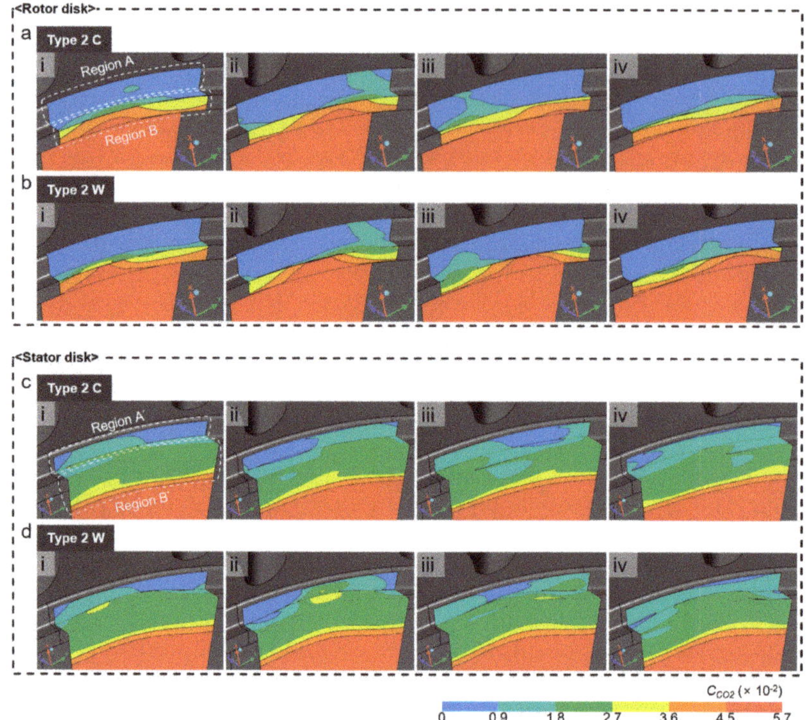

Figure 11. Contour of C_{CO2} on the surface of the rotor disk of (**a**) Type 2 C and (**b**) Type 2 W, and the stator disk of (**c**) Type 2 C and (**d**) Type 2 W.

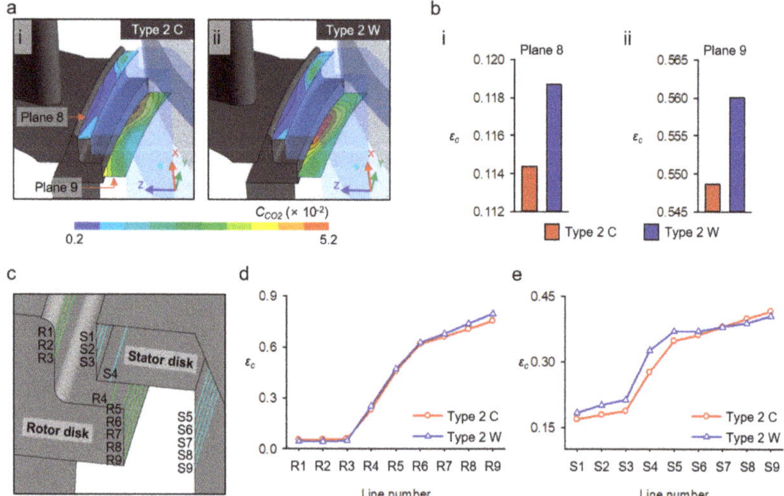

Figure 12. (**a**) Contour of C_{CO2} in Plane 8 near the mainstream path, and Plane 9 located adjacent to the rotor radial seal in the wheelspace; (**i**) Type 2 C and (**ii**) Type 2 W. (**b**) Comparison of sealing effectiveness between Type 2 C and Type 2 W in (**i**) Plane 8 and (**ii**) Plane 9. (**c**) Position of R1–9 on the rotor disk and S1–9 on the stator disk. Sealing effectiveness in (**d**) R1–9 and (**e**) S1–9.

4. Conclusions

In this computational study, a wave-shaped rim seal geometry was proposed for application in stator radial seals. Four rim seal geometries were analysed via CFD to observe the positive effect of the wave-shaped rim seal of the stator side. Two-step sequences of computations were performed; first, to isolate the wave-shape geometry effect on the local flow dynamic change. Type 1 models, in which the stator vane and rotor blade were not included, were simulated to observe the flow around the radial seal of conventional and wave-shape cases. Then, Type 2 models, which include the stator vane and rotor blade, were simulated to show how the wave-shaped rim seal improves its sealing performance. In Type 1 W, flow with high pressure and low velocity was found in the ridge region of the wave-shaped rim seal where the radial gap is maximised, and the faster flow at lower pressure was found in the trough region, where the radial gap is minimised. This pressure distribution near the radial seal of Type 1 W has a similar tendency to the pressure distribution of the mainstream flow (shown in Figure 1b(ii)) formed owing to the wake of the stator vane in the mainstream path of the actual turbine. When the wave-shaped rim seal, which caused the flow phenomenon observed in Type 1 W, was applied to the geometry with the rotor blades and stator vanes (Type 2 W), the radial velocity, i.e., the egress and ingress flow, decreased compared to Type 2 C owing to the change in pressure distribution near the wave-shaped radial seal in the interface between the mainstream path and wheelspace up to 1.12% and 2.67% in the egress and ingress flow, respectively. The C_{CO2} contour shows that the wave-shaped rim seal geometry induces the coolant flow to better protect the rotor and stator disks and wheelspace. Compared with Type 2 C, Type 2 W has a 5.7% and 17% increase in sealing effectiveness for the rotor and stator disks, respectively.

The proposed wave-shaped rim seal and its positive resultant effect indicate that three-dimensional deformations of the stator radial seal can further improve the sealing effectiveness to control pressure distribution around the rim seal, which is ultimately the overall turbine efficiency. This type of new geometric parameter is useful for future gas turbine design. Further studies on optimising the wave geometric parameters will have better results and a greater impact on the improvement of rim seals.

Supplementary Materials: The following are available online at http://www.mdpi.com/1996-1073/12/6/1060/s1.

Author Contributions: Conceptualization, S.L. and J.Y.P.; Methodology, S.L. and J.Y.P.; Software, S.L. and D.K.; Formal Analysis, S.L. and D.K.; Writing-Original Draft Preparation, S.L.; Writing-Review & Editing, S.L. and J.Y.P.; Supervision, J.Y.P.

Funding: This research was supported by the Korea Institute of Energy Technology Evaluation and Planning (KETEP) funded by the Ministry of Trade, Industry and Energy, Republic of Korea (No.20163010024690), and also supported by the Chung-Ang University Graduate Research Scholarship in 2018 (D.H.K.).

Acknowledgments: The authors would like to thank Senior Researcher Young Sang Kim from Doosan Heavy Industries and Construction for his helpful advice on various aspects of the gas turbine design process.

Conflicts of Interest: The authors declare no conflict of interest.

Nomenclature

C_{CO2}	Mass fraction of CO_2
C_c	Coolant flow inlet mass fraction of CO_2
C_h	Mainstream inlet mass fraction of CO_2
C_l	Local mass fraction of CO_2
P_{abs}	Absolute pressure, Pa
P_{static}	Static pressure, Pa
$P_{ave,static}$	Area-averaged static pressure, Pa
R	Disk radius, m
r	Radial distance from rotation axis, m
V_{cir}	Circumferential velocity, m/s
$V_{ave,cir}$	Area-averaged circumferential velocity, m/s
V_{eg}	Radial velocity component of egress flow, m/s

V_{in}	Radial velocity component of ingress flow, m/s
$V_{max,eg}$	Maximum radial velocity component of egress flow, m/s
$V_{min,in}$	Minimum radial velocity component of ingress flow, m/s
V_{rad}	Radial velocity, m/s
ε_c	Sealing effectiveness
ζ	Normalised angle, θ/θ_p
θ	Angle, °
θ_p	Periodic angle, °

References

1. Razak, A.M.Y. *Industrial Gas Turbines: Performance and Operability*; Elsevier: Amsterdam, The Netherlands, 2007; ISBN 184569340X.
2. Owen, J.M. Prediction of Ingestion Through Turbine Rim Seals—Part II: Externally Induced and Combined Ingress. *J. Turbomach.* **2011**, *133*, 31006. [CrossRef]
3. Batchelor, G.K. Note on a class of solutions of the Navier-Stokes equations representing steady rotationally-symmetric flow. *Q. J. Mech. Appl. Math.* **1951**, *4*, 29–41. [CrossRef]
4. Stewartson, K. On the flow between two rotating coaxial disks. In *Proceedings of the Mathematical Proceedings of the Cambridge Philosophical Society*; Cambridge University Press: Cambridge, UK, 1953; Volume 49, pp. 333–341.
5. Poncet, S.; Chauve, M.-P.; Schiestel, R. Batchelor versus Stewartson flow structures in a rotor-stator cavity with throughflow. *Phys. Fluids* **2005**, *17*, 75110. [CrossRef]
6. Owen, J.M.; Pountney, O.; Lock, G. Prediction of Ingress Through Turbine Rim Seals—Part II: Combined Ingress. *J. Turbomach.* **2012**, *134*, 31013. [CrossRef]
7. Roy, R.P.; Zhou, D.W.; Ganesan, S.; Wang, C.-Z.; Paolillo, R.E.; Johnson, B.V. The flow field and main gas ingestion in a rotor-stator cavity. In Proceedings of the ASME Turbo Expo 2007: Power for Land, Sea, and Air, Montreal, QC, Canada, 14–17 May 2007; American Society of Mechanical Engineers: New York, NY, USA; pp. 1189–1198.
8. Zhou, D.W.; Roy, R.P.; Wang, C.-Z.; Glahn, J.A. Main gas ingestion in a turbine stage for three rim cavity configurations. *J. Turbomach.* **2011**, *133*, 31023. [CrossRef]
9. Bohn, D.; Wolff, M. *Improved Formulation to Determine Minimum Sealing Flow—Cw, min—For Different Sealing Configurations*; ASME Pap. No. GT2003-38465; The American Society of Mechanical Engineers: New York, NY, USA, 2003.
10. Eastwood, D.; Coren, D.D.; Long, C.A.; Atkins, N.R.; Childs, P.R.N.; Scanlon, T.J.; Guijarro-Valencia, A. Experimental Investigation of Turbine Stator Well Rim Seal, Re-Ingestion and Interstage Seal Flows Using Gas Concentration Techniques and Displacement Measurements. *J. Eng. Gas Turbines Power* **2012**, *134*, 82501. [CrossRef]
11. Teuber, R.; Li, Y.S.; Maltson, J.; Wilson, M.; Lock, G.D.; Owen, J.M. Computational extrapolation of turbine sealing effectiveness from test rig to engine conditions. *Proc. Inst. Mech. Eng. Part A J. Power Energy* **2013**, *227*, 167–178. [CrossRef]
12. Wang, C.-Z.; Mathiyalagan, S.P.; Johnson, B.V.; Glahn, J.A.; Cloud, D.F. Rim Seal Ingestion in a Turbine Stage From 360 Degree Time-Dependent Numerical Simulations. *J. Turbomach.* **2014**, *136*, 31007. [CrossRef]
13. Phadke, U.P.; Owen, J.M. Aerodynamic aspects of the sealing of gas-turbine rotor-stator systems: Part 2: The performance of simple seals in a quasi-axisymmetric external flow. *Int. J. Heat Fluid Flow* **1988**, *9*, 106–112. [CrossRef]
14. Phadke, U.P.; Owen, J.M. Aerodynamic Aspects of the Sealing of Gas-Turbine Rotor-Stator Systems: Part 3: The Effect of Nonaxisymmetric External Flow on Seal Performance. *Int. J. Heat Fluid Flow* **1988**, *9*, 113–117. [CrossRef]
15. Bunker, R.S.; Laskowski, G.M.; Bailey, J.C.; Palafox, P.; Kapetanovic, S.; Itzel, G.M.; Sullivan, M.A.; Farrell, T.R. An Investigation of Turbine Wheelspace Cooling Flow Interactions With a Transonic Hot Gas Path—Part 1: Experimental Measurements. *J. Turbomach.* **2011**, *133*, 21015. [CrossRef]

16. Palafox, P.; Ding, Z.; Bailey, J.; Vanduser, T.; Kirtley, K.; Moore, K.; Chupp, R. *A New 1.5-Stage Turbine Wheelspace Hot Gas Ingestion Rig (HGIR)—Part I: Experimental Test Vehicle, Measurement Capability and Baseline Results*; ASME Pap. No. GT2013-96020; The American Society of Mechanical Engineers: New York, NY, USA, 2013.
17. Sangan, C.M.; Pountney, O.J.; Scobie, J.A.; Wilson, M.; Owen, J.M.; Lock, G.D. Experimental Measurements of Ingestion Through Turbine Rim Seals—Part III: Single and Double Seals. *J. Turbomach.* **2013**, *135*, 51011. [CrossRef]
18. Popovíc, I.; Hodson, H.P. Improving turbine stage efficiency and sealing effectiveness through modifications of the rim seal geometry. *J. Turbomach.* **2013**, *135*, 61016. [CrossRef]
19. Moon, M.-A.; Lee, C.-S.; Kim, K.-Y. Effect of a rib on rim seal performance. *Int. Commun. Heat Mass Transf.* **2014**, *59*, 130–135. [CrossRef]
20. Scobie, J.A.; Teuber, R.; Li, Y.S.; Sangan, C.M.; Wilson, M.; Lock, G.D. Design of an improved turbine rim-seal. *J. Eng. Gas Turbines Power* **2016**, *138*, 22503. [CrossRef]
21. Sangan, C.M.; Scobie, J.A.; Owen, J.M.; Lock, G.D.; Tham, K.M.; Laurello, V.P. Performance of a finned turbine rim seal. *J. Turbomach.* **2014**, *136*, 111008. [CrossRef]
22. Liu, D.; Tao, Z.; Luo, X.; Wu, H.; Yu, X. Development of a new factor for hot gas ingestion through rim seal. *J. Eng. Gas Turbines Power* **2016**, *138*, 72501. [CrossRef]
23. Li, J.; Gao, Q.; Li, Z.; Feng, Z. Numerical Investigations on the Sealing Effectiveness of Turbine Honeycomb Radial Rim Seal. *J. Eng. Gas Turbines Power* **2016**, *138*, 102601. [CrossRef]
24. Balasubramanian, J.H. *Experimental Study of Main Gas Ingestion and Purge Gas Egress Flow in Model Gas Turbine Stages*; Arizona State University: Tempe, AZ, USA, 2010; ISBN 1124359184.

Article

Gas Turbine Cycle with External Combustion Chamber for Prosumer and Distributed Energy Systems

Dariusz Mikielewicz [1], Krzysztof Kosowski [1,*], Karol Tucki [2,*], Marian Piwowarski [1], Robert Stępień [1], Olga Orynycz [3,*] and Wojciech Włodarski [1]

1 Faculty of Mechanical Engineering, Gdansk University of Technology, Gabriela Narutowicza Street 11/12, 80-233 Gdansk, Poland; dariusz.mikielewicz@pg.edu.pl (D.M.); marian.piwowarski@pg.edu.pl (M.P.); rstepien@pg.edu.pl (R.S.); wwlodar@pg.edu.pl (W.W.)

2 Department of Organization and Production Engineering, Warsaw University of Life Sciences, Nowoursynowska Street 164, 02-787 Warsaw, Poland

3 Department of Production Management, Bialystok University of Technology, Wiejska Street 45A, 15-351 Bialystok, Poland

* Correspondence: kosowski@pg.gda.pl (K.K.); karol_tucki@sggw.pl (K.T.); o.orynycz@pb.edu.pl (O.O.); Tel.: +48-593-45-78 (K.T.); +48-746-98-40 (O.O.)

Received: 16 July 2019; Accepted: 10 September 2019; Published: 11 September 2019

Abstract: The use of various biofuels, usually of relatively small Lower Heating Value (LHV), affects the gas turbine efficiency. The present paper shows that applying the proposed air by-pass system of the combustor at the turbine exit causes tan increase of efficiency of the turbine cycle increased by a few points. This solution appears very promising also in combined gas/steam turbine power plants. The comparison of a turbine set operating according to an open cycle with partial bypassing of external combustion chamber at the turbine exit (a new solution) and, for comparison, a turbine set operating according to an open cycle with a regenerator. The calculations were carried out for different fuels: gas from biomass gasification (LHV = 4.4 MJ/kg), biogas (LHV = 17.5 MJ/kg) and methane (LHV = 50 MJ/kg). It is demonstrated that analyzed solution enables construction of several kW power microturbines that might be used on a local scale. Such turbines, operated by prosumer's type of organizations may change the efficiency of electricity generation on a country-wide scale evidently contributing to the sustainability of power generation, as well as the economy as a whole.

Keywords: biofuels; sustainable power generation; microturbines

1. Introduction

An increase in the demand for energy and its carriers, as well as the growing legal requirements in the aspect of harmful emissions cause [1–3] the necessity to continuously increase the share of energy from renewable sources [4–6]. In addition to energy from water, wind or solar radiation, biomass is the most commonly used source of renewable fuel [7–9]. Its importance as an inexpensive and renewable source of energy grows particularly when it is produced from agricultural and forestry residues [10,11]. The use of by-products or residues of the agri-food industry for biogas or biofuel production is frequently observed. The growing level of public awareness together with the tightening of legal requirements in environmental protection issues add dynamics to the increase in the level of attractiveness of activities and technologies leading to the reduction of adverse human impact on the ecosystem [12–14]. The sustainability of biomass derived energetics based on distributed electricity generation can be assessed by means of life cycle assessment (LCA) methodology, as well as the direct approach offered by [15–17]. During managing of the production process, the amount of energy consumed and the production of post-production waste emitted to the atmosphere are analyzed at each stage of the production process. The analysis of energy expenditure is carried out in a comprehensive manner, concerning the whole life cycle of the product. In this context, even electromobility requires a huge amount of energy used to extract and process lithium, cobalt, and manganese; the key raw materials needed to produce batteries for electric cars. Thermodynamic cycles and the use of modern calculation tools are also important in this process [18,19].

The development of prosumer energetics, is expected to assure strengthening of energy security, to implement climate policy, but also create the opportunity to solve grid problems and also to mitigate an increase of energy prices [20–22]. The observed dynamic changes in the structure of the electricity mix are caused, among others, by changes in the development strategy of individual countries, which are motivated by principles of sustainable development and, as a consequence, climate policy being an indispensable element of planning reforms in the power sector [23]. This leads to a visible increase in the number of prosumer generation centers being built (also biogas production plants/biorefineries with polygeneration devices) [24,25]. In these types of units, most often turbines, attention is paid to durability, reliability, low price or efficiency of components and the entire installation [26,27].

Biogas plants and biorefineries are, apart from technology, connected to either dedicated energy crops (biogas-biomethane) or to lignocellulosic wastes of plants used for edible crops (second generation biofuels-biomethanol) [28–30]. Biogas obtained in the process of anaerobic methane fermentation can be considered as a substitute for natural gas and a universal source of cheap energy produced and used locally [31,32]. Production of biomethane is, in turn, the clean for the environment, and the fuel is characterized by a higher EROEI index than vegetable oils [33]. The division of biofuels into individual generations is based mainly on the type of raw material processed [34]. Currently, R&D activities in biorefineries are aimed at the bioconversion of lignocellulose to simple sugars and fermentation to ethanol. In addition, the work focuses on breeding new varieties of energy plants with high biomass efficiency (e.g., 2nd generation bioethanol from hemp mass). Third generation biofuels (products obtained as a result of the conversion of new raw materials intended for this purpose) seem to be still a distant future [35–37]. The only enterprise on the wider scale operating in Poland that attempts to produce third generation biofuels is PKN Orlen [38,39]. The experimental station launched in Płock, in which the technology for the production of biocomponents derived from oil algae is being developed [40]. For their production, post-production water and CO_2 obtained from refining processes will be used [41,42].

The innovativeness of modern biogas plants is primarily connected with new technologies for the production of energy crops and models of "tapes" of plant production and preservation assuring the continuous supply of energetic biomass along with the necessary logistic security [43,44]. In addition, innovative processes for the fermentation are used due to the microbiological composition of the fermentation flora depending on the biomass feedstock parameters. R&D works are conducted towards the optimization of the biochemical processes of the bioreactor depending on the type of plant biomass and the use of biogas in the fuel cell [45,46]. The disadvantage of the domestic power generation sector is the relatively low efficiency of energy production from coal, and in the case of dispersed power engineering, the efficiencies of small power plants are even lower, and thermal power plants based on circuits with organic agents achieve efficiencies of only a dozen or so percent [47–50]. In addition, there is the issue of high carbon dioxide emissions [51].

As part of the electric micro-plants' development, one can indicate turbine micro-turbines [52,53], bladeless-free adhesive turbines [54] or solar collectors [55,56]. Turbines of small output can be arranged in simple or complex cycles, including regenerators, interstage coolers or successive combustion chambers. Turbines burning biofuels in external combustion chambers arouse particular interest. In this solution, clean air flows in compressor and turbine during the whole period of operation. In the case of an external combustion chamber, it is possible to bleed some turbine exhaust air and omit a combustion chamber. This can improve turbine unit efficiency.

A key role in the modern cogeneration units belongs to the high efficiency and (in parallel) compact heat exchangers with passive techniques of heat transfer augmentation. Preferential are the heat exchangers with minichannels of cylindrical construction [57] and plate ones [58]. Very promising are also the heat exchangers with the micro-jets' technology—very intensive experimental and numerical investigations of their development are conducted at the moment.

Particular biofuels can differ depending on their chemical composition and the heating values which influence the relations between the temperatures and the mass (and volume) flow rates of the working media. This, in turn, shows some impact on the power plant overall efficiency and the design of turbomachinery flow parts. In the case of various biofuels (especially pellets) so called "external combustion systems" may be used, which allows to burn different sorts of fuel (liquid, gas or solid), even of poor quality, because in these units' clean air flows through the compressor and the turbine.

Currently, it is possible to obtain a stable flame during the combustion of low calorific fuels in a wide range of operating parameters, such as the molar composition of the fuel and the excess air coefficient. However, the biogas must be properly cleaned and dried so that it does not damage the turbine. Depending on the origin, the biogas composition is variable. The calorific value depends primarily on the methane content. Currently, biogas, that is combusted in gas turbines, has the methane content from 35% to 100%. As a result of continuous combustion with excess air and low pressures in the combustion chamber, turbines as well as microturbines have a significantly lower value of exhaust emissions as compared to the reciprocating engines. The combustion of low calorific gases has a significant impact on the natural environment by reducing the emission of nitrogen oxides [59,60].

If the fuel of low value of Lower Heating Value (LHV) is used in gas turbine units, the mass flow rate of a turbine exhaust air can be remarkably higher than mass flow rate of compressor air. Thus, in the case of an external combustion chamber, it is possible to bleed some turbine exhaust air and omit a combustion chamber. This original arrangement was compared with the open gas turbine cycle with regenerator. Thus, two variants were considered during the calculations:

- Variant 1: turbine set operating according to the open cycle with regenerator (Figure 1),
- Variant 2: turbine set operating according to the open cycle with partial bypassing of external combustion chamber at turbine exit and with high-temperature heat exchanger (Figure 2).

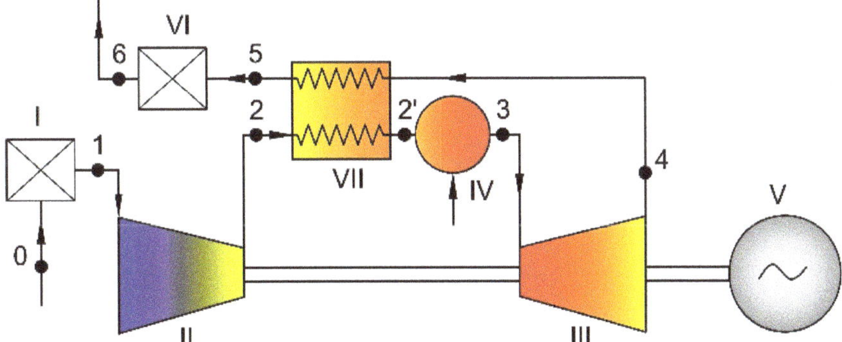

Figure 1. Turbine set operating according to the open cycle with regenerator (Variant 1-V1). I–filter; II–compressor; III–turbine; IV–combustion chamber; V–electric generator; VI–silencer; VII–regenerator.

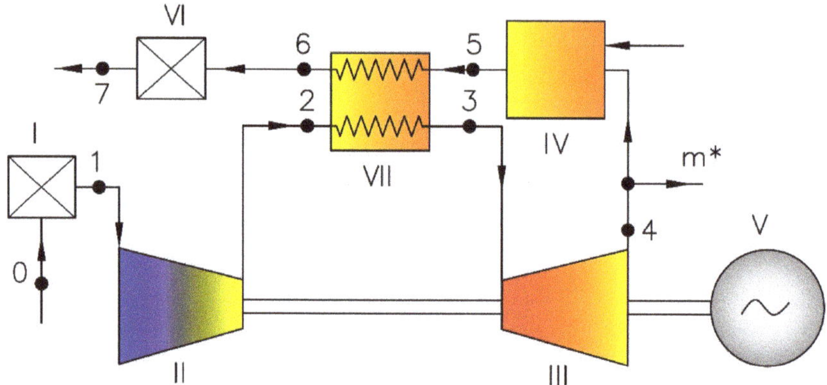

Figure 2. Turbine set operating according to the open cycle with partial bypassing of the external combustion chamber at the turbine exit and with a high-temperature heat exchanger (Variant 2 -V2). I–filter; II–compressor; III–turbine; IV–external combustion chamber; V–electric generator; VI–silencer; VII–regenerator.

2. Materials and Methods

2.1. Nomentclature and Units

The following list contains the most important symbols and units (Tables 1 and 2) used in the formulae and in the figures.

Table 1. Symbols and units used in calculations.

Symbol	Description	Unit
c_p	specific heat at constant pressure	[kJ/kg·K]
h	enthalpy of unit mass	[kJ/kg]
p	pressure	[Pa]
η	efficiency	[–]
$\dot{m}$	mass flow	[kg/s]
l	work of unit mass	[kJ/kg]
c_p	specific heat at constant pressure	[kJ/kg·K]
κ	exponent of isentropic process	[–]
W	power	[kW]
$\dot{Q}$	heat flux	[kW]
T	temperature	[·K]
LHV	lower heating value	$[MJ/m^3]$ or [MJ/kg]
Π	compression ratio	[–]

Table 2. List of used subscripts.

Symbol	Description
f	fuel
m	mechanical
opt	optimal
meth	methane
C	compressor
CC	combustion chamber
T	turbine
GT	gas turbine set
1, 2, ...	points on diagrams

2.2. The Computation Algorithm

In the case of small power plants (from several kW to several hundred kW), the maximum temperature 900 °C was assumed before the turbine, and the low efficiency of the components was assumed. Assumptions adopted for the analysis are presented in Table 3.

Table 3. Assumptions adopted for the design analysis of turbine generator variants [27,51].

Description	Unit	Value
compressor efficiency	[-]	0.800
turbine efficiency	[-]	0.820
mechanical efficiency	[-]	0.980
leakage losses	[-]	0.02
generator efficiency	[-]	0.900
efficiency of the combustion chamber	[-]	0.950
pressure loss in the filter	[-]	0.995
pressure loss in the silencer	[-]	0.995
pressure losses in the combustion chamber	[-]	0.98
pressure loss in the regenerator	[-]	0.98

The results of thermodynamical calculations of the cycles with gas turbines using fuels of different caloric values were presented in the paper. The results are presented for a gas from biomass gasification and for a biogas. For comparison, the analysis was also carried out for methane (the main component of LNG or natural gas). The characteristics of the considered gases are shown in Table 4.

Table 4. The list of analyzed gases with different heating values [61,62].

Fuel Type	Volumetric Composition	Density	Calorific Value, LHV
	[-]	[kg/m^3]	[MJ/kg]
gas from biomass gasification	CH_4 0.09; CO_2 0.133; CO 0.147; H_2 0.073; N_2 0.42; H_2O 0.137	1.2107	4.4
biogas	CH_4 0.4; CO_2 0.23; H_2 0.16; CO 0.1; N_2 0.11	0.9438	17.5
methane	CH_4 1.0	0.6660	54.1

The thermodynamical calculations were performed following classical approach known from the bibliography [27,63–65]. The calculations were performed in the following order: the calculations of compression process in the compressor, calculations of expansion line in the turbine, calculations of regenerator and combustion chamber energy balance equations. The following main relationships were used in the analysis, including energy balances and definitions for the efficiencies.

The power and specific work of the gas turbine set are determined accordingly to the dependencies:

$$W_{GT} = \eta_m \cdot \dot{m}_T \cdot l_T - \dot{m}_C \cdot l_C$$

$$l_{GT} = \frac{W_{GT}}{\dot{m}_C} = \eta_m \cdot \left(\frac{\dot{m}_T}{\dot{m}_C}\right) \cdot l_T - \left(\frac{\dot{m}_C}{\dot{m}_C}\right) \cdot l_C = \eta_m \cdot \left(\frac{\dot{m}_T}{\dot{m}_C}\right) \cdot l_T - l_C$$

The specific work of the compressor and turbine are defined by relations:

$$l_C = \frac{1}{\eta_C} \cdot c_{pC} \cdot T_1 \cdot \left(\left(\frac{p_2}{p_1}\right)^{\frac{\kappa_C - 1}{\kappa_C}} - 1 \right)$$

$$l_T = \eta_T \cdot c_{pT} \cdot T_3 \left(1 - \left(\frac{p_4}{p_3}\right)^{\frac{\kappa_T - 1}{\kappa_T}} \right)$$

The efficiency of the gas turbine cycle is defined as:

$$\eta_{GT} = \frac{W_{GT}}{\dot{Q}_1}$$

where the heat flux brought to combustion chamber:

$$\dot{Q}_1 = \left(\dot{m}_T \cdot h_3 - \dot{m}_C \cdot h_{2'}\right)\frac{1}{\eta_{CC}} = \dot{m}_f \cdot LHV$$

After the transformations of the above equations, it was obtained:

$$\eta_{GT} = \frac{\eta_m \cdot \dot{m}_T \cdot l_T - \dot{m}_C \cdot l_C}{\left(\dot{m}_T \cdot i_3 - \dot{m}_C \cdot i_{2'}\right) \cdot \frac{1}{\eta_{CC}}}$$

and:

$$\eta_{TG} = \eta_{KS} \cdot \frac{\eta_m \cdot \eta_C \cdot c_{pT} \cdot \frac{T_3}{T_1}\left[1 - \left(\frac{1}{\Pi_T}\right)^{\frac{\kappa_T - 1}{\kappa_T}}\right] \cdot \frac{\dot{m}_T}{\dot{m}_C} - \frac{1}{\eta_C} \cdot c_{pC} \cdot \left[\left(\Pi_C\right)^{\frac{\kappa_C - 1}{\kappa_C}} - 1\right]}{\frac{\dot{m}_T}{\dot{m}_C} \cdot c_{pCC} \cdot \frac{T_3}{T_1} - \frac{1}{\eta_C} \cdot c_{pC} \cdot \left[\left(\Pi_C\right)^{\frac{\kappa_C - 1}{\kappa_C}} - 1\right]}$$

The heat flux transferred from the exhaust fumes in the regenerator is that:

$$\dot{Q}_{VII,\,T} = \dot{m}_T \cdot (h_4 - h_5)$$

The heat flux received by the air in the regenerator is that:

$$\dot{Q}_{VII,\,C} = \dot{m}_C \cdot (h_{2'} - h_2)$$

Average values of specific heat for particular states of working media were determined on the basis of their chemical composition and thermodynamical parameters using REFPROP software.

3. Results

Calculations of two variants of cycles were performed (Figures 1 and 2) for three fuels with different lower heating values (Table 4). The reference was made to the results for methane as a comparative fuel. For each of the analyzed variants, the compression ratio was optimized to obtain maximum efficiency. The effect of the compression on the value of the relative efficiency (referred to the maximum value) for three fuels used in variant 1 is shown in Figure 3, and for variant 2-in Figure 4. As can be seen, the calorific value shows an effect on the optimal compression (maximum efficiency) only for variant 1, whereas in variant 2 the type of fuel has a little effect on the optimal compression value. This is confirmed by the results shown in Figure 5 (optimal compression values for different types of fuel and both variants).

The Figure 6, in turn, shows the influence of the structure of the cycle and calorific value of the fuel on the relative efficiency (referred to methane) for both variants of turbine sets. In each case considered, the decrease of calorific value leads to a reduction in the efficiency of the turbine set (from a few to a dozen or so percent). The subsequent figures show the influence of the cycle structure and of calorific value of the fuel on the relative mass flow of fuel combusted in the combustion chamber (Figure 7), on the relative mass flow of exhaust (Figure 8) and on the relative power of the turbine set all related to the corresponding variants of the cycles with methane used as a fuel (Figure 9).

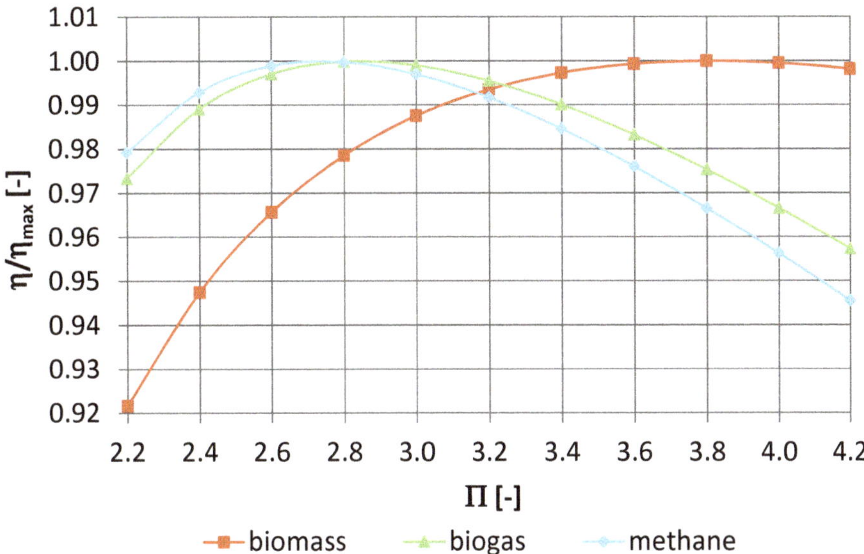

Figure 3. Effect of compression ratio ∏ on the value of relative efficiency (referred to the maximum value) for three example fuels in variant 1.

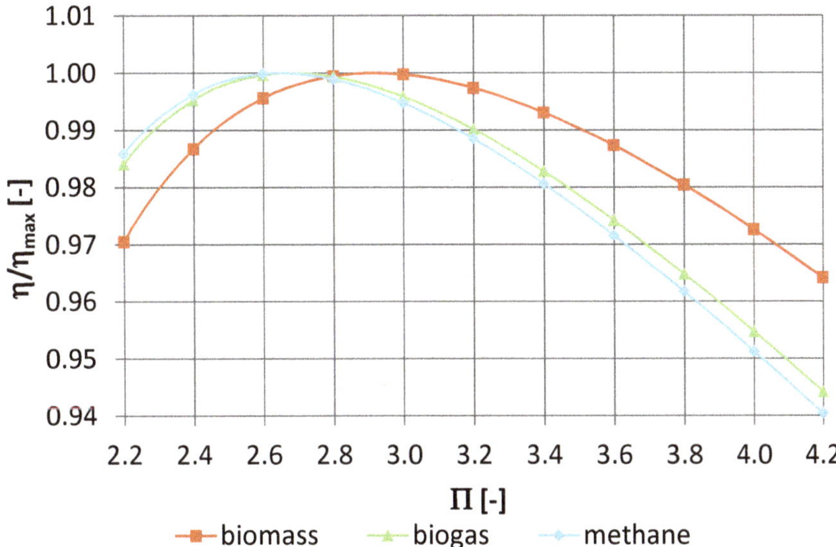

Figure 4. Effect of compression ratio ∏ on the value of relative efficiency (referred to the maximum value) for three exemplary fuels in variant 2.

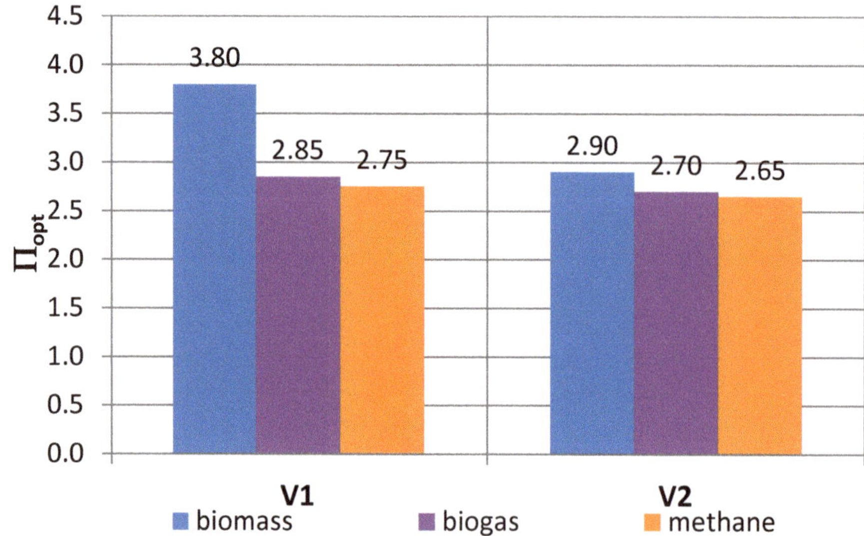

Figure 5. Influence of the structure of the cycle and calorific value of fuel on the optimal compression.

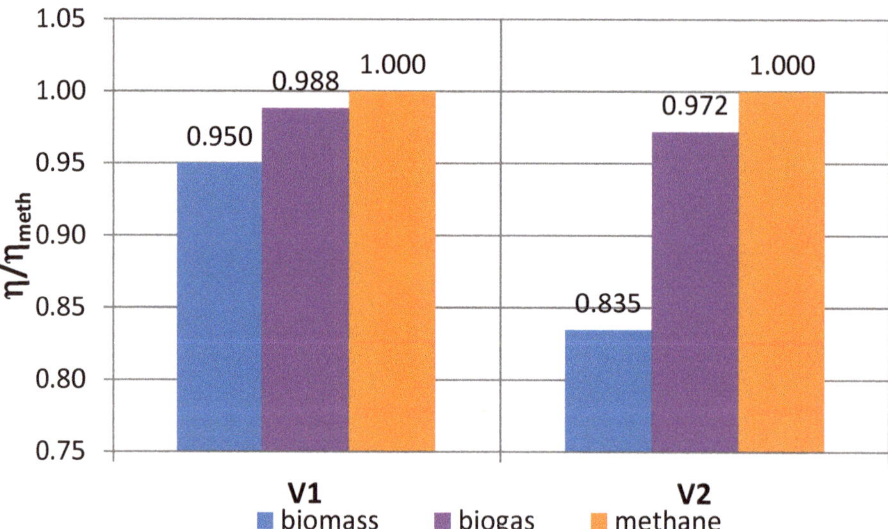

Figure 6. The influence of the structure of the cycle and calorific value of the fuel on the relative efficiency (referred to methane) for both variants of turbine sets.

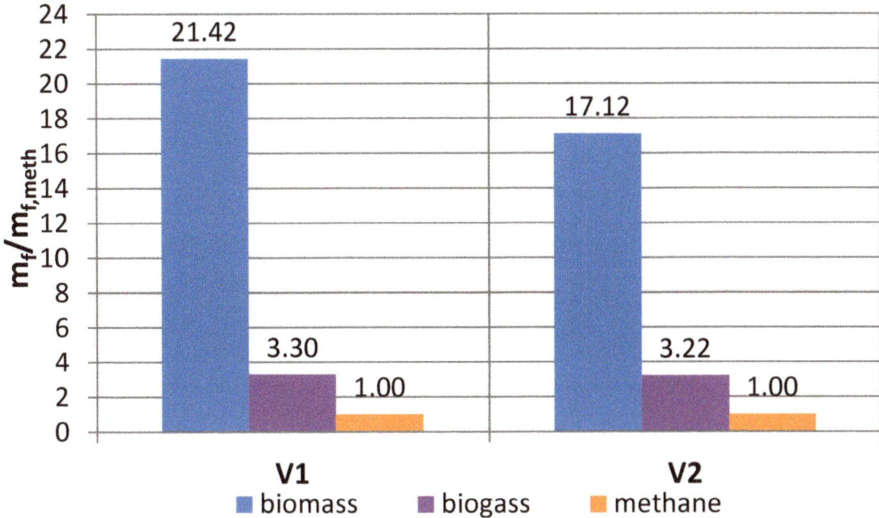

Figure 7. The influence of the structure of the cycle and calorific value of fuel on the relative mass stream of the fuel in the combustion chamber for both variants (referred to methane as a fuel).

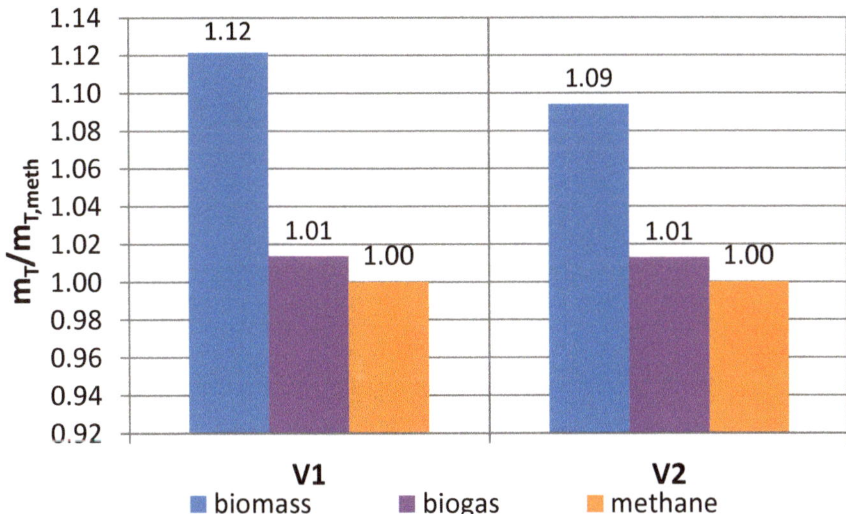

Figure 8. Effect of the structure of the cycle and calorific value of fuel on the exhaust gas mass flow for both variants (related to methane used as a fuel).

A decrease of the lower heating value leads to a clear, multiple increase in mass fuel consumption, especially when using low calorific biomass gasification fuel (Figure 7). However, due to the small mass fraction of fuels in relation to air, this does not significantly affect the increase of exhaust gas flow (Figure 8). Only in the case of fuel from biomass, the increase of the flue gas stream may amount to approx. 10% and only in variant 1 this results in a very large increase in the unit power of the turbine set.

It is interesting to compare directly the efficiency of both variants of the turbine set cycles (V1 and V2). Only for biomass fuel, the variant with the external combustion chamber (V2) appears worse than the classic turbine set with the regenerator (V1). For higher caloricity of the fuel, variant 2 achieves higher efficiency than variant 1, Figure 10.

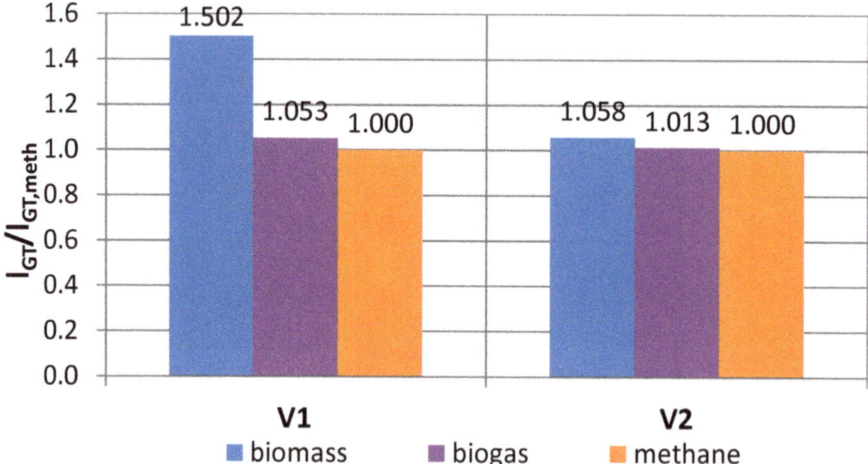

Figure 9. The influence of the structure of the cycle and calorific value of fuel for both variants on the effective power related to that with methane as a fuel.

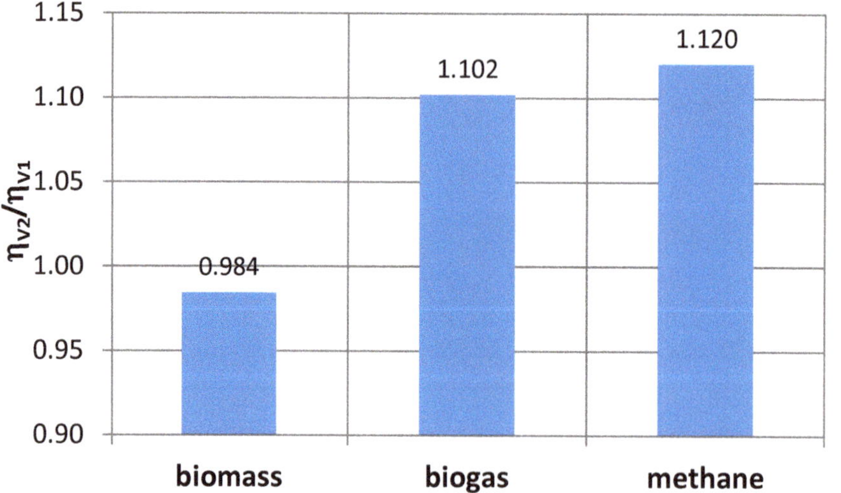

Figure 10. The influence of calorific value of the fuel on the efficiency ratio of the cycles with the external combustion chamber and bypass (V2), and for with the regenerator (V1).

In the analysis of the work of cycles, one should also consider the possibilities of using energy of exhaust gas in cogeneration systems for generating electricity and heat as well as in combined systems (cooperation with other types of thermal cycles). Therefore, in this analysis the values of the stream intensity and the temperature of the medium leaving the turbine set are important. Figure 11 shows the outlet air temperature (after the regenerator) for both variants V1 and V2, and for all the fuels considered. The highest outlet temperatures are for biomass fuel (with the lowest caloricity LHV) for both the regenerator cycle (278 °C) and the cycle with the external combustion chamber (160 °C).

Cycles with an external combustion chamber are usually characterized by a lower outlet temperature of the working medium.

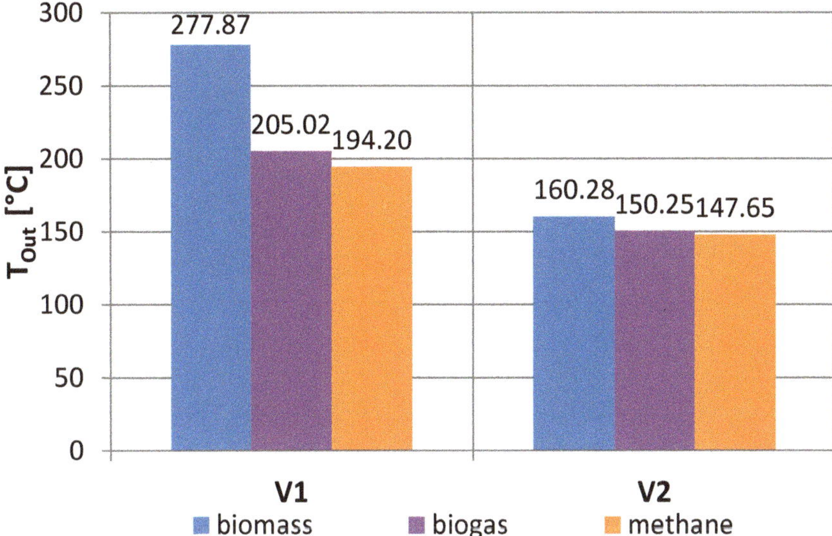

Figure 11. The influence of the calorific value of the fuel and the structure of the cycle on temperature of exhaust gas leaving the turbine set.

Figure 12 shows the relative air flow drawn from behind the turbine in relation to the air flow through the compressor. It ranges from a few to a dozen or so percent (for the fuel from biomass), but it should be taken into account that it is a high temperature air, which creates greater opportunities for its use, also in combined systems. The possibility of heat production in cogeneration with electricity generation for variants with a regenerator and an external combustion chamber is shown in Figure 13 (variant 3) and Figure 14 (variant 4), respectively.

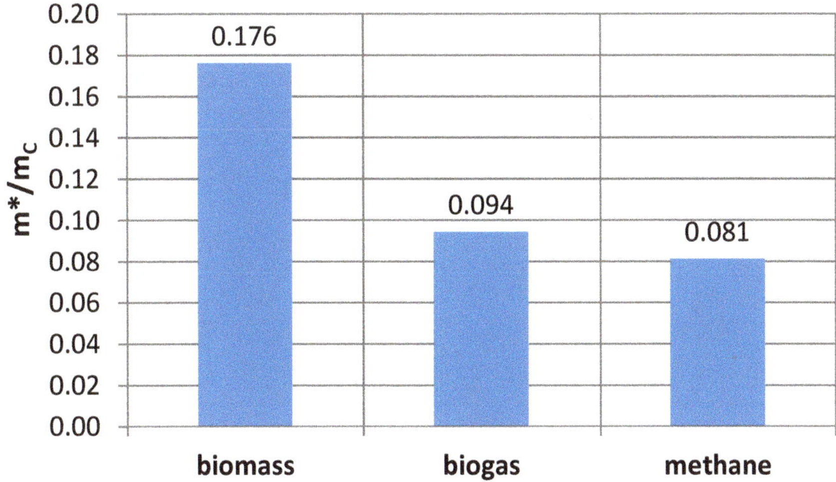

Figure 12. The influence of the calorific value of the fuel on the relative mass stream of the bypass to the mass flow of air flowing through the compressor for the optimal compression ratio.

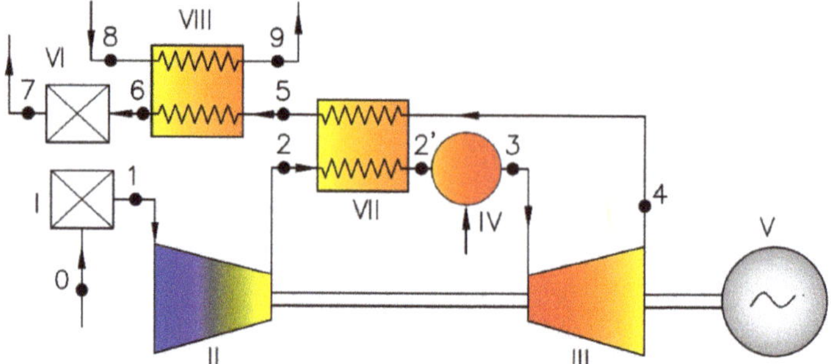

Figure 13. Circulation diagram with a gas turbine with a regenerator and cogeneration (Variant 3-V3). I–filter; II–compressor; III–turbine; IV–combustion chamber; V–electric generator; VI–silencer; VII–regenerator; VIII- heat exchanger (for cogeneration heat).

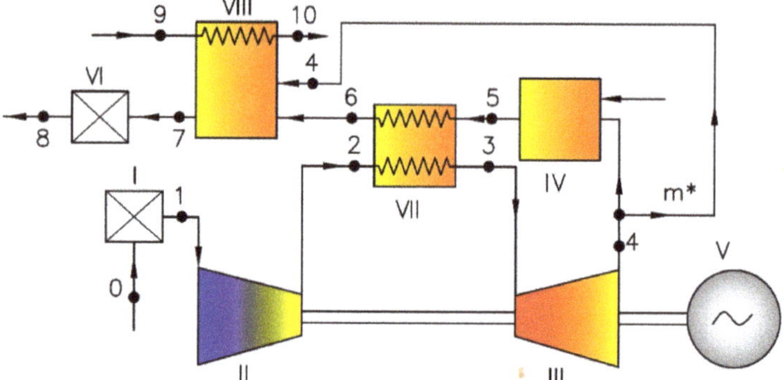

Figure 14. Circulation diagram with air turbine, external combustion chamber and bypass and cogeneration (Variant 4-V4). I–filter; II–compressor; III–turbine; IV–external combustion chamber; V–electric generator; VI–silencer; VII–regenerator; VIII- heat exchanger (for cogeneration heat).

Figure 15 shows a gas turbine with an external combustion chamber and air intake from behind the turbine for combined cycle with a steam turbine (variant 5). This solution increases the electric power generated, and the efficiency of the entire power plant. The comparison of the efficiency of the combined machine (variant 5) with the efficiency of the system with regenerator (variant 1) is shown in Figure 16. Regardless of the fuel used, the efficiency of the combined system (variant 5) exceeded about 30% efficiency of turbine set with regenerator (variant 1).

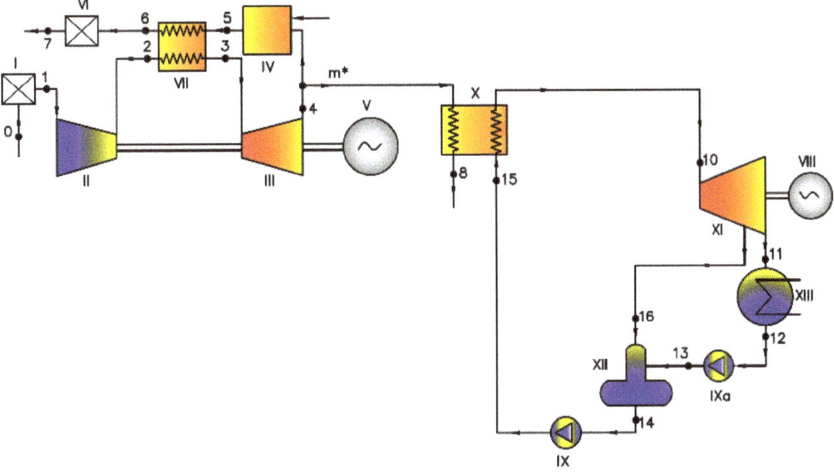

Figure 15. Circulation diagram with an air turbine, an external combustion chamber and a bypass with an additional steam circuit (Variant 5-V5). I–filter; II–compressor; III–turbine; IV–external combustion chamber; V–electric generator of gas turbine; VI–silencer; VII–regenerator; VIII- electric generator of steam turbine; IX–main pump; IXa–condensate pump; X–heat recovery steam generator; X1–steam turbine; XII–direct contact heat exchanger; XIII–condenser.

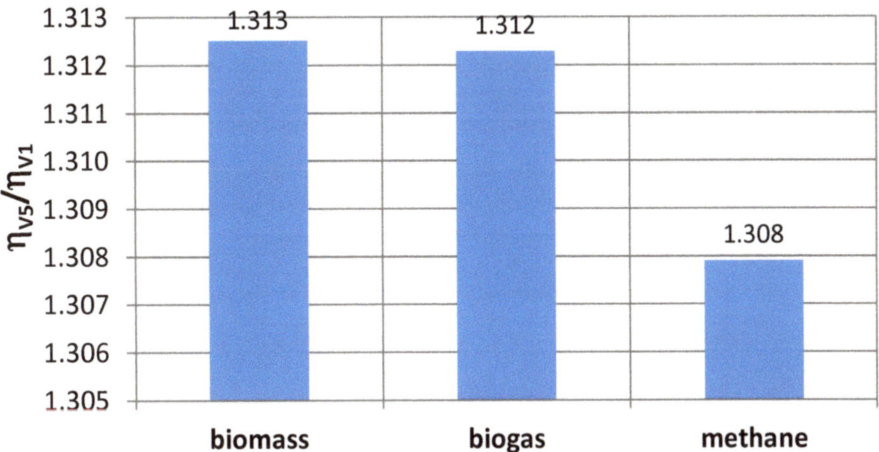

Figure 16. The effect of calorific value of fuel on the relative efficiency of the cycle with the external combustion chamber and bypass and additional steam circuit (V5) referred to the cycle with regenerator (V1).

The results of calculations for the analyzed fuels are presented in Tables 5 and 6.

Table 5. Summary of the results of calculations carried out for the different calorific values.

Parameter	Variant	Biomass	Biogas	Methane
Π_{opt}	V1	3.80	2.85	2.75
	V2	2.90	2.70	2.65
η/η_{meth}	V1	0.950	0.988	1.000
	V2	0.835	0.972	1.000
$m_f/m_{f,meth}$	V1	21.42	3.30	1.00
	V2	17.12	3.22	1.00
$m_T/m_{T,meth}$	V1	1.12	1.01	1.00
	V2	1.09	1.01	1.00
$l_{GT}/l_{GT,meth}$	V1	1.502	1.053	1.000
	V2	1.058	1.013	1.000
T_{out}	V1	277.87	205.02	194.20
	V2	160.28	150.25	147.65

Table 6. Summary of the results of calculations carried out for the different calorific values of fuel for selected cycles.

Fuel	η_{V2}/η_{V1}	m^*/m_c
Biomass	0.984	0.176
Biogas	1.102	0.094
Methane	1.120	0.081

All the calculations were performed assuming efficiencies of the main turbine power plants elements. Efficiencies of compressors, turbines, combustion chambers and regenerators (Table 3) can be treated as quite good values for gas turbine sets of small output. If we consider the other values of these parameters, we may expect a change of overall power plant efficiency, for example:

- applying lower compressor efficiency equal to 0.75 (instead 0.80) we observe the drop of overall efficiency by about 7%–8% for all considered variants while assuming higher compressor efficiency 0.85 the increase in overall efficiency can reach 6.5%–7.5%,
- decreasing turbine efficiency from 0.82 to 0.77 we lower the overall efficiency by about 9%–10% and increasing it to 0.87 we have the rise in overall efficiency by about 8%–9%,
- changing the combustion chamber efficiency by 5% the overall efficiency also changes by about 5%,
- assuming high values of all elements (compressor efficiency −0.85, turbine efficiency −0.87, combustion chamber efficiency −0.985, regenerator efficiency 0.98) we may increase the overall power plant efficiency by 18%–21%,
- using low values of all elements (compressor efficiency −0.75, turbine efficiency −0.77, combustion chamber efficiency −0.9, regenerator efficiency 0.8) we may reduce the overall power plant efficiency by 28%–32%.

All the considered variants, in spite of the fuel used, are more sensitive to reduction than to growth of the efficiency of particular power plant elements.

For many years, installations with electric power of several hundred kilowatts and more have been used in biomass-burning power plants. Cogeneration systems working with organic media are already available in a wide range of electrical and thermal power, e.g., a power plant with an electrical power of 300–600 kW and a heat power of 1500–2800 kW or a heat and power plant with an electric power of 200–1000 kW and a thermal power of 1000–6000 kW [66]. However, only a few examples of Organic Rankin Cycle (ORC) installations with an output power less than 100 kW [67] or less than 30 kW [68] of electrical power can be found. Only a few examples of ORC cogeneration systems

(Organic Rankin Cycle) with an electric power below 5 kW can be found on the market. The published results of experimental studies on the operation of microturbine sets can be found, for example, in [69].

Previous research has shown that it is possible to build a set of microturbines with a capacity of about 2 kW with a higher efficiency than in existing machines [70]. It is worth noting that the relatively high efficiency of microturbines can be achieved thanks to a very careful and advanced design process (Figure 17). The safe and reliable behavior of the microturbine set has been confirmed during operational tests. The results indicate the need to consider the interaction between components of the ORC installation and microturbine.

(a)

(b)

Figure 17. Example of experimental single stage micro turbine. (**a**) view of experimental micro turbine; (**b**) bladed rotor of micro turbine with shaft and electrical generator rotor.

4. Final Conclusions

The performed analysis have shown that:

1. Systems with an external combustion chamber and air intake from behind the turbine can be competitive in terms of efficiency with classic turbine sets, also those with regeneration.
2. The use of fuels with lower calorific value usually leads to a lower efficiency of the power plant.
3. The structure of the gas turbine power plant and the type of fuel should be taken into account at the stage of determining the main design parameters of the turbine set (e.g., optimal compression ratio, unit power).
4. Turbosets with an external combustion chamber and air intake from behind the turbine are characterized by wide possibilities of applications in cogeneration systems and high efficiency combined systems.

Conducted analyses provide knowledge to help to mitigate potential environmental hazards through introduction of biofuels into distributed energy generation and optimization of turbines to such, locally available fuels.

Energy efficiency solutions provide opportunities for diversification and reduce energy consumption as well as primary consumption. It seems that the efficiency of turbine sets for micro-energetics will clearly increase after the introduction of new high-efficiency thermodynamic cycles. Such effective technological changes and improvements to the system may be one of the stages in the fight against the energy crisis, as well as an element of strategic development in the shaping of the national power generation sector.

The problem still requires further research, but implementation of the final might contribute to the reduction of environmental burdens.

At present two experimental micro power plants with external combustion chambers are being built: a domestic cogeneration power plant of 20 kW heat power and 2 kW electric power as well as a high efficiency turbine power plant with an innovative isothermal turbine of 5 kW (Figure 18).

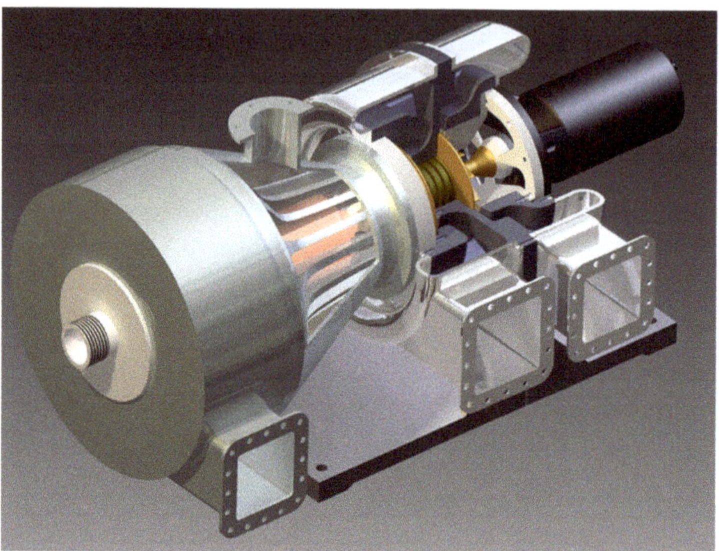

Figure 18. Example of an experimental variant of the innovative isothermal turbine.

Author Contributions: Conceptualization, K.K. and M.P.; R.S.; W.W.; Methodology, K.T. and O.O.; D.M.; Investigation, R.S. and W.W.; Writing—Original Draft Preparation, K.T.; M.P. and O.O.; Funding Acquisition, K.K. and D.M.

Funding: The Authors wish to express their deep gratitude to Gdansk University of Technology for financial support given to the present publication (Krzysztof Kosowski). The research was carried out under financial support obtained from the research subsidy of the Faculty of Engineering Management (WIZ) of Bialystok University of Technology (Olga Orynycz).

Conflicts of Interest: The authors declare no conflict of interest. The funders had no role in the design of the study; in the collection, analyses, or interpretation of data; in the writing of the manuscript, and in the decision to publish the results.

References

1. Maczynska, J.; Krzywonos, M.; Kupczyk, A.; Tucki, K.; Sikora, M.; Pinkowska, H.; Baczyk, A.; Wielewska, I. Production and use of biofuels for transport in Poland and Brazil. The case of bioethanol. *Fuel* **2019**, *241*, 989–996. [CrossRef]
2. Tucki, K.; Mruk, M.; Orynycz, O.; Botwinska, K.; Gola, A. Toxicity of Exhaust Fumes (CO, NOx) of the Compression-Ignition (Diesel) Engine with the Use of Simulation. *Sustainability* **2019**, *11*, 2188. [CrossRef]
3. Tucki, K.; Mruk, R.; Orynycz, O.; Wasiak, A.; Botwinska, K.; Gola, A. Simulation of the Operation of a Spark Ignition Engine Fueled with Various Biofuels and Its Contribution to Technology Management. *Sustainability* **2019**, *11*, 2799. [CrossRef]
4. Braungardt, S.; Bürger, V.; Zieger, J.; Bosselaar, L. How to include cooling in the EU Renewable Energy Directive? Strategies and policy implications. *Energy Policy* **2019**, *129*, 260–267. [CrossRef]
5. Veum, K.; Bauknecht, D. How to reach the EU renewables target by 2030? An analysis of the governance framework. *Energy Policy* **2019**, *127*, 299–307. [CrossRef]
6. Gökgöz, F.; Güvercin, M.T. Energy security and renewable energy efficiency in EU. *Renew. Sustain. Energy Rev.* **2018**, *96*, 226–239. [CrossRef]

7. Xu, G.; Li, M.; Lu, P. Experimental investigation on flow properties of different biomass and torrefied biomass powders. *Biomass Bioenergy* **2019**, *122*, 63–75. [CrossRef]
8. Goffé, J.; Ferrasse, J.H. Stoichiometry impact on the optimum efficiency of biomass conversion to biofuels. *Energy* **2019**, *170*, 438–458.
9. Tucki, K.; Orynycz, O.; Wasiak, A.; Swic, A.; Wichlacz, J. The Impact of Fuel Type on the Output Parameters of a New Biofuel Burner. *Energies* **2019**, *12*, 1383. [CrossRef]
10. Bunn, D.W.; Redondo-Martin, J.; Munoz-Hernandez, J.I.; Diaz-Cachinero, P. Analysis of coal conversion to biomass as a transitional technology. *Renew. Energy* **2019**, *132*, 752–760. [CrossRef]
11. Hamelin, L.; Borzęcka, M.; Kozak, M.; Pudełko, R. A spatial approach to bioeconomy: Quantifying the residual biomass potential in the EU-27. *Renew. Sustain. Energy Rev.* **2019**, *100*, 127–142. [CrossRef]
12. Holland, R.A.; Beaumont, N.; Hooper, T.; Austen, M.; Gross, R.J.K.; Heptonstall, P.J.; Ketsopoulou, I.; Winskel, M.; Watson, J.; Taylor, G. Incorporating ecosystem services into the design of future energy systems. *Appl. Energy* **2018**, *222*, 812–822. [CrossRef]
13. Picchi, P.; Van Lierop, M.; Geneletti, D.; Stremke, S. Advancing the relationship between renewable energy and ecosystem services for landscape planning and design: A literature review. *Ecosyst. Serv.* **2019**, *35*, 241–259. [CrossRef]
14. Tucki, K.; Mruk, R.; Orynycz, O.; Gola, A. The Effects of Pressure and Temperature on the Process of Auto-Ignition and Combustion of Rape Oil and Its Mixtures. *Sustainability* **2019**, *11*, 3451. [CrossRef]
15. Wasiak, A.; Orynycz, O. The effects of energy contributions into subsidiary processes on energetic efficiency of biomass plantation supplying biofuel production system. *Agric. Agric. Sci. Procedia* **2015**, *7*, 292–300. [CrossRef]
16. Moslehi, S.; Reddy, T.A. An LCA methodology to assess location-specific environmental externalities of integrated energy systems. *Sustain. Cities Soc.* **2019**, *46*, 1–14. [CrossRef]
17. Kupczyk, A.; Maczynska, J.; Redlarski, G.; Tucki, K.; Baczyk, A.; Rutkowski, D. Selected Aspects of Biofuels Market and the Electromobility Development in Poland: Current Trends and Forecasting Changes. *Appl. Sci.* **2019**, *9*, 254. [CrossRef]
18. Kosowski, K.; Tucki, K.; Kosowski, A. Application of Artificial Neural Networks in Investigations of Steam Turbine Cascades. *J. Turbomach. Trans. ASME* **2010**, *132*, 014501–014505. [CrossRef]
19. Kosowski, K.; Tucki, K.; Kosowski, A. Turbine stage design aided by artificial intelligence methods. *Expert Syst. Appl.* **2009**, *36*, 11536–11542. [CrossRef]
20. IEO Report PV Market in Poland 2019. Available online: www.ieo.en/ (accessed on 17 July 2019).
21. Tucki, K.; Orynycz, O.; Wasiak, A.; Świć, A.; Dybaś, W. Capacity market implementation in Poland: Analysis of a survey on consequences for the electricity market and for energy management. *Energies* **2019**, *12*, 839. [CrossRef]
22. Baatz, B.; Relf, G.; Nowak, S. The role of energy efficiency in a distributed energy future. *Electr. J.* **2018**, *31*, 13–16. [CrossRef]
23. Puri, S.; Perera, A.T.D.; Mauree, D.; Coccolo, S.; Delannoy, L.; Scartezzini, J.L. The role of distributed energy systems in European energy transition. *Energy Procedia* **2019**, *159*, 286–291. [CrossRef]
24. Calise, F.; De Notaristefani di Vastogirardi, G.; D'Accadia, M.D.; Vicidomini, M. Simulation of polygeneration systems. *Energy* **2018**, *163*, 290–337. [CrossRef]
25. Jarnut, M.; Wermiński, S.; Waśkowicz, B. Comparative analysis of selected energy storage technologies for prosumer-owned microgrids. *Renew. Sustain. Energy Rev.* **2017**, *74*, 925–937. [CrossRef]
26. Aiying, R.; Risto, L. Role of polygeneration in sustainable energy system development: Challenges and opportunities from optimization viewpoints. *Renew. Sustain. Energy Rev.* **2016**, *53*, 363–372.
27. Kosowski, K.; Domachowski, Z.; Próchnicki, W.; Kosowski, A.; Stępień, R.; Piwowarski, M.; Włodarski, W.; Ghaemi, M.; Tucki, K.; Gardzilewicz, A.; et al. *Steam and Gas Turbines with the Examples of Alstom Technology*, 1st ed.; Alstom: Saint-Quen, France, 2007; ISBN 978-83-925959-3-9.
28. Krzywonos, M.; Tucki, K.; Wojdalski, J.; Kupczyk, A.; Sikora, M. Analysis of Properties of Synthetic Hydrocarbons Produced Using the ETG Method and Selected Conventional Biofuels Made in Poland in the Context of Environmental Effects Achieved. *Rocz. Ochr. Śr.* **2017**, *19*, 394–410.
29. Tonini, D.; Hamelin, L.; Alvarado-Morales, M.; Astrup, T.F. GHG emission factors for bioelectricity, biomethane, and bioethanol quantified for 24 biomass substrates with consequential life-cycle assessment. *Bioresour. Technol.* **2016**, *208*, 123–133. [CrossRef]

30. Shen, X.; Kommalapati, R.R.; Huque, Z. The Comparative Life Cycle Assessment of Power Generation from Lignocellulosic Biomass. *Sustainability* **2015**, *7*, 12974. [CrossRef]
31. Carneiro, M.L.N.M.; Pradelle, F.; Braga, S.L.; Gomes, M.S.P.; Martins, M.R.F.A.; Turkovics, F.; Pradelle, R.N.C. Potential of biofuels from algae: Comparison with fossil fuels, ethanol and biodiesel in Europe and Brazil through life cycle assessment (LCA). *Renew. Sustain. Energy Rev.* **2017**, *73*, 632–653. [CrossRef]
32. Adelt, M.; Wolf, D.; Vogel, A. LCA of biomethane. *J. Nat. Gas Sci. Eng.* **2011**, *3*, 646–650. [CrossRef]
33. Ardolino, F.; Arena, U. Biowaste-to-Biomethane: An LCA study on biogas and syngas roads. *Waste Manag.* **2019**, *87*, 441–453. [CrossRef] [PubMed]
34. Rocha, M.H.; Capaz, R.S.; Lora, E.E.S.; Nogueira, L.A.H.; Leme, M.M.V.; Renó, M.L.G.; Del Olmo, O.A. Life cycle assessment (LCA) for biofuels in Brazilian conditions: A meta-analysis. *Renew. Sustain. Energy Rev.* **2014**, *37*, 435–459. [CrossRef]
35. Collotta, M.; Champagne, P.; Mabee, W.; Tomasoni, G.; Leite, G.B.; Busi, L.; Alberti, M. Comparative LCA of Flocculation for the Harvesting of Microalgae for Biofuels Production. *Procedia CIRP* **2017**, *61*, 756–760. [CrossRef]
36. Milledge, J.J.; Heaven, S. Energy Balance of Biogas Production from Microalgae: Effect of Harvesting Method, Multiple Raceways, Scale of Plant and Combined Heat and Power Generation. *J. Mar. Sci. Eng.* **2017**, *5*, 9. [CrossRef]
37. Gonzalez-Fernandez, C.; Sialve, B.; Bernet, N.; Steyer, J.P. Impact of microalgae characteristics on their conversion to biofuel. Part ii: Focus on biomethane production. *Biofuels Bioprod. Biorefining* **2012**, *6*, 205–218. [CrossRef]
38. Igliński, B.; Iglińska, A.; Kujawski, W.; Buczkowski, R.; Cichosz, M. Bioenergy in Poland. *Renew. Sustain. Energy Rev.* **2011**, *15*, 2999–3007. [CrossRef]
39. Experimental Algae to Grow at the Refinery. Available online: https://www.orlen.pl/ (accessed on 17 July 2019).
40. PKN ORLEN Invests in Biofuels of the Future. Available online: https://www.orlen.pl/ (accessed on 17 July 2019).
41. ORLEN Group 2017. Integrated Report. Available online: https://raportzintegrowany2017.orlen.pl/ (accessed on 17 July 2019).
42. Unipetrol Enters the BIOENERGY 2020+ project. Available online: http://www.unipetrol.cz (accessed on 17 July 2019).
43. Lisowski, A. *Technologie Zbioru Roślin Energetycznych (Energy Plants Harvesting Technologies)*, 1st ed.; Wydawnictwo SGGW: Warszawa, Polska, 2010; pp. 77–119.
44. Stolarski, M.J.; Niksa, D.; Krzyżaniak, M.; Tworkowski, J.; Szczukowski, S. Willow productivity from small- and large-scale experimental plantations in Poland from 2000 to 2017. *Renew. Sustain. Energy Rev.* **2019**, *101*, 461–475. [CrossRef]
45. Cozzolino, R.; Lombardi, L.; Tribioli, L. Use of biogas from biowaste in a solid oxide fuel cell stack: Application to an off-grid power plant. *Renew. Energy* **2017**, *111*, 781–791. [CrossRef]
46. Saadabadi, S.A.; Thattai, A.T.; Fan, L.; Lindeboom, R.E.F.; Spanjers, H.; Aravind, P.V. Solid Oxide Fuel Cells fuelled with biogas: Potential and constraints. *Renew. Energy* **2019**, *134*, 194–214. [CrossRef]
47. Mikielewicz, J.; Piwowarski, M.; Kosowski, K. Design analysis of turbines for co-generating micro-power plant working in accordance with organic rankine's cycle. *Pol. Marit. Res.* **2009**, *1*, 34–38. [CrossRef]
48. Stępniak, D.; Piwowarski, M. Analyzing selection of low-temperature medium for cogeneration micro power plant. *Pol. J. Environ. Stud.* **2014**, *23*, 1417–1421.
49. Tucki, K.; Sikora, M. Technical and logistics analysis of the extension of the energy supply system with the cogeneration unit supplied with biogas from the water treatment plant. *TEKA of the Commission of Motorization and Power Industry in Agriculture* **2016**, *16*, 71–75.
50. Piwowarski, M.; Kosowski, K. Design analysis of combined gas-vapour micro power plant with 30 kw air turbine. *Pol. J. Environ. Stud.* **2014**, *23*, 1397–1401.
51. Kosowski, K.; Tucki, K.; Piwowarski, M.; Stępień, R.; Orynycz, O.; Włodarski, W.; Bączyk, A. Thermodynamic Cycle Concepts for High-Efficiency Power Plans. Part A: Public Power Plants 60+. *Sustainability* **2019**, *11*, 554. [CrossRef]
52. Giorgetti, S.; Parente, A.; Bricteux, L.; Contino, F.; De Paepe, W. Optimal design and operating strategy of a carbon-clean micro gas turbine for combined heat and power applications. *International Journal of Greenhouse Gas Control* **2019**, *88*, 469–481. [CrossRef]

53. De Paepe, W.; Coppitters, D.; Abraham, S.; Tsirikoglou, P.; Ghorbaniasl, G.; Contino, F. Robust Operational Optimization of a Typical micro Gas Turbine. *Energy Procedia* **2019**, *158*, 5795–5803. [CrossRef]
54. Lampart, P.; Kosowski, K.; Piwowarski, M.; Jędrzejewski, L. Design analysis of tesla micro-turbine operating on a low-boiling medium. *Pol. Marit. Res.* **2009**, *1*, 28–33. [CrossRef]
55. Kabir, E.; Kumar, P.; Kumar, S.; Adelodun, A.A.; Kim, K.H. Solar energy: Potential and future prospects. *Renew. Sustain. Energy Rev.* **2018**, *82*, 894–900. [CrossRef]
56. Palm, J.; Eidenskog, M.; Luthander, R. Sufficiency, change, and flexibility: Critically examining the energy consumption profiles of solar PV prosumers in Sweden. *Energy Res. Soc. Sci.* **2018**, *39*, 12–18. [CrossRef]
57. Chordia, L. High temperature heat exchanger design and fabrication for systems with large pressure differentials. In *Final Scientific/Technical Report 2017*. Available online: www.osti.gov/servlets/purl/1349235 (accessed on 17 July 2019).
58. The exceptional performance of Heatric PCHE heat exchangers. Available online: www.heatric.com/heat_exchanger_performance.html (accessed on 17 July 2019).
59. Ślefarski, R.; Jójka, J.; Czyżewski, P.; Grzymisławski, P. Experimental investigation on syngas reburning process in a gaseous fuel firing semi-industrial combustion chamber. *Fuel* **2018**, *217*, 490–498. [CrossRef]
60. Włodarski, W. Control of a vapour microturbine set in cogeneration applications. *ISA Transa.* **2019**, in press. [CrossRef]
61. Taler, J.; Mruk, A.; Cisek, J.; Majewski, K. Combined heat and power plant with internal combustion engine fuelled by wood gas. *Rynek Energii* **2013**, *4*, 62–67.
62. Kordylewski, W. *Spalanie i Paliwa*, 5th ed.; Oficyna Wydawnicza Politechniki Wrocławskiej: Wrocław, Polska, 2008; pp. 10–470. ISBN 978-83-7493-378-0.
63. Traupel, W. *Thermische Turbomachinen*, 2nd ed.; Springer: Berlin, Heidelberg; New York, NY, USA, 1982; pp. 122–525. Available online: https://link.springer.com/book/10.1007%2F978-3-642-96632-3 (accessed on 12 August 2019).
64. Sawyer, J.W. *Gas Turbine Engineering Handbook*, 3rd ed.; Turbomachinery International Publications: Norwalk, CT, USA, 1985; pp. 98–245. ISBN 0-937506-14-1.
65. Sorensen, H.A. *Gas Turbines (Series in Mechanical Engineering)*, 1st ed.; Ronald Press Company: New York, NY, USA, 1951; pp. 48–440.
66. Gailfuß, M. *Private meets Public—Small Scale CHP*; Technological Developments, Workshop BHKW-Infozentrum Rastatt 09.09.2003: Berlin, Germany, 2003.
67. Kosowski, K.; Piwowarski, M.; Stępień, R.; Włodarski, W. Design and Investigations of a Micro-Turbine Flow Part. In Proceedings of the ASME Turbo Expo 2012: Turbine Technical Conference and Exposition; Volume 5: Manufacturing Materials and Metallurgy; Marine; Microturbines and Small Turbomachinery; Supercritical CO_2 Power Cycles, Copenhagen, Denmark, 11–15 June 2012; Paper No. GT2012-69222. pp. 807–814. [CrossRef]
68. Mills, D. Advances in solar thermal electricity technology. *Solar Energy* **2004**, *76*, 19–31. [CrossRef]
69. Kosowski, K.; Włodarski, W.; Piwowarski, M.; Stepień, R. Performance characteristics of a micro-turbine. *Adv. Vib. Eng.* **2014**, *2*, 341–350.
70. Kosowski, K.; Piwowarski, M.; Stępień, R.; Włodarski, W. Design and investigations of the ethanol microturbine. *Arch. Thermodyn.* **2018**, *39*, 41–54.

Article

Computational Simulation of PT6A Gas Turbine Engine Operating with Different Blends of Biodiesel—A Transient-Response Analysis

Camilo Bayona-Roa *, J.S. Solís-Chaves, Javier Bonilla, A.G. Rodriguez-Melendez and Diego Castellanos

Universidad ECCI, Cra. 19 No. 49-20, Bogotá 111311, Colombia; jsolisc@ecci.edu.co (J.S.S.-C.); jbonillap@ecci.edu.co (J.B.); rodriguez.andersong@ecci.edu.co (A.G.R.-M.); diegocastellanos1234@hotmail.com (D.C.)

* Correspondence: cbayonar@ecci.edu.co

Received: 17 September 2019; Accepted: 8 October 2019; Published: 8 November 2019

Abstract: Instead of simplified steady-state models, with modern computers, one can solve the complete aero-thermodynamics happening in gas turbine engines. In the present article, we describe a mathematical model and numerical procedure to represent the transient response of a PT6A gas turbine engine operating at off-design conditions. The aero-thermal model consists of a set of algebraic and ordinary differential equations that arise from the application of the mass, linear momentum, angular momentum and energy balances in each engine's component. The solution code has been developed in Matlab-Simulink® using a block-oriented approach. Transient simulations of the PT6A engine start-up have been carried out by changing the original Jet-A1 fuel with biodiesel blends. Time plots of the main thermodynamic variables are shown, especially those regarding the structural integrity of the burner. Numerical results have been validated against reported experimental measurements and GasTurb® simulations. The computer model has been capable to predict acceptable fuel blends, such that the real PT6A engine can be substituted to avoid the risk of damaging it.

Keywords: gas turbine engine; two-spool turboprop engine; PT6A engine; aero-thermal model; Matlab-Simulink; bio-diesel; start-up transient

---

1. Introduction

The necessary thrust that is required for an aircraft to provide lift is commonly supplied by a heat engine. In particular, a gas turbine engine can convert heat energy into mechanical energy by involving the flow of air passing through several thermo-fluidic processes within its components. One of the most important features of gas turbine engines is that, contrary to reciprocating engines, separate sections of the engine are devoted to the intake, compression, combustion, power conversion and exhaust processes. This also means that all processes are performed simultaneously and that they are strongly coupled between them. Hence, the conceptual description of the gas turbine engine's response is complex; it involves the approximation from very different engineering disciplines: Aerodynamics, Thermodynamics, Heat Transfer, Structural Analysis, Materials Science and Mechanical Design.

Remarkably, in all those fields the mathematical modeling of a gas turbine engine is constructed by applying the mass, energy and momentum balances over the most representative solid and fluid elements of the engine. These balances have to do with the air crossing through the gas turbine engine's stages and its interaction with the rotatory solid components. The overall result of this modeling process gives a set of several partial differential equations whose solution is the spatial and temporal description of the air through the engine's stages and the performance of the engine. Since there are not any exact solutions due to the geometry complexity and temporal dependence, we rely entirely on numerical methods to find an approximated numerical solution.

The level of numerical resolution of each component is also important: a great level of detail in the description may also be computationally expensive. Selecting one or other may be based on the objectives of the simulation and the given computational resources. In a turboprop engine modeling, for example, the complex geometric description of the rotary components makes it hard to generate a full spatial solution, namely a discrete spatial mesh for Computational Fluid Dynamics (CFD). Considering the dynamic rotatory movement of the components raises—even more—the model complexity and the computational cost. Trying to couple the engine's components in the CFD simulation is a problem that may not be solvable with actual computational methods and power.

And that is our particular motivation in the present work: to simulate the overall engine performance. We are also restricted by the computational resources at hand, which are a personal computer. These requirements typically lead to neglect a very detailed spatial description of the airflow, preferring an integrated definition of each engine stage to what is referred to as a lower-dimensional model. This approach defines space-averaged values at each engine's component and the partial differential equations arising from the balance equations turn to ordinary differential equations that can be solved numerically in an affordable way. The important aspect is that this spatially-simplified approach is capable to represent the engine's operation giving accurate physical solutions that arise from the balance equations.

Precisely, we are interested in computational simulations capable of representing the PT6A turboprop gas turbine engine performance when a change in the operating fuel occurs. Specifically, the evaluation of the transient state of the engine at the start-up procedure, when the temperature in the burner reaches its maximum and can significantly affect the combustion chamber. Testing a gas turbine operating with fuel compositions is costly and time-consuming but also leads to performance degradation of the manufactured engine. Instead, the tests that can be achieved with computational models like those in References [1–4] which are inexpensive and fast when used for new designs, predicting the system's integrity and calculating real-time responses. In this sense, a wide range of computer engines have been implemented and reported in the literature: from systematic models in References [5–9], to the detailed component design of engine parts in References [10–15]. Because of the previously supported reasons, we opt for an integrated representation of the engine as an assembly of components to what is commonly referred to as a Common Engineering Model (CEM). These components—inlet, fan, compressors, combustion, turbine, shafts and nozzle—are abstracted in the computational model as blocks—or objects—, having connections between them and therefore, being able to model the various propulsion systems, such as multi-spool or turbo-fan type of engines.

We restrict our survey to engine models based on physical descriptions of the engine components. Those range from simulation tools used by the gas turbine production and control industry [16], steady and transient performance simulations of power gas turbines in References [17,18] and general gas turbine software such as DYNGEN [19], TERTS [20], GSP [21], Gasturbolib [22], GasTurb® [23], among others. Academic research has exploited the architectures given by Matlab-Simulink® [22] and Modelica® [24], where different engine types can be created in a visual interface approach. However, most of these works simulate the steady-state thermodynamic cycle of the engine without describing the dynamic behavior of the air that crosses the stages of the engine and nearly all are linearized versions that do not work far from the linearization point —like when the engine operates with blends of biodiesel fuel. Instead, the coupled non-linear version of the equations are solved with the Newton-Raphson method but it requires the computation of the Jacobian and may not converge to the actual numerical solution at all operating regimes. The particular methods that we are interested in, are the ones that do not require an extensive iterative process to obtain the numerical solution of the set of algebraic and differential equations. Those arrange the equations in such a way that the output values can be obtained from the input variables. This is consistent with the flow of air through the engine components: the input variables are referred to the entering flow conditions, while the output solved variables are the exiting flow.

The thermal performance prediction of gas turbine engines operating with biodiesel has barely been reported in References [25,26] using these computational techniques. The opposite is true for reciprocating engines operating with biodiesel blends, for which performances and emissions predictions have been extensively carried out. What was reported for the steady simulations of gas turbine engines indicated that the modified lower heating value of biodiesel fuels had a significant influence on thrust, fuel flow, and specific fuel consumption at every flight condition and all mixing ratio percentages. Those preliminary results are encouraging and motivate the present research in gas turbine engine technology. Chiefly, for substituting the source of jet fuel from fossil-based fuel to biomass-based, which could reduce the environmental impact and energy consumption of gas turbines.

Our computational approach for the transient response simulation of a gas turbine engine operating with blends of biodiesel is similar to the ones in References [7,20,22,27], where aero-thermodynamical descriptions of gas turbine engines have been done. In contrast to those approaches, we consider the gas dynamics, which are the application of the mass, momentum and energy balances at every component of the gas turbine engine. Especially at the compression and turbine sections, where a stage-by-stage thermo-fluidic description is done for high-fidelity purposes. Another key aspect of the present work has to do with the solution algorithms and the development of the model in a flexible programming architecture, which makes it possible to create the specific configuration of the PT6A turboprop engine but also enables the future extension of the engine model. We formulate each engine component using a separate block, in which the mass, momentum and energy equations give the solution of the dynamical variables. The complete engine is the composition of blocks resulting in a block diagram. Matlab-Simulink® offers the capability of coding those blocks and connecting them to configure any gas turbine engine. It also gives the possibility of using a graphical interface which is more user-friendly, but the relevant feature of this software is that it makes possible to generate source code in C language to generate a much faster executable program. Block diagrams make also possible to link the engine software to control devices, with the capability of driving input/output signals for data acquisition and control. All these features being relevant in the on-live prediction of the gas turbine engine.

The remaining parts of this article are organized in the following order. In Section 2, the real operational gas turbine engine is described: we reproduce a generic version of the Pratt-Whitney PT6A engine. Simultaneously, we present the theoretical description of the thermodynamic processes occurring inside each engine's components. The simulation of different scenarios are given in Section 3. The scenarios include the validation of our model with the reported experimental data of the engine's steady operation, the subsequent steady operation with blends of biodiesel and the transient engine response at the start-up procedure to determine the engine's structural integrity and deterioration. Finally, in Section 4 some conclusions close the article.

2. The PT6A Engine Model

A generic version of the Pratt-Whitney PT6A-65 "Medium" engine, including the labeling of its parts, is displayed in Figure 1. In general, gas turbine engines consist of: an air inlet, a compression section (including the compressor and diffuser), a combustion section (combustion chamber), a turbine section, an exhaust section and an accessory section (for necessary systems). The theoretical description of the thermodynamic processes occurring in these components is Brayton's cycle. But the dynamic simulation of the engine's behavior must account for non-equilibrium conditions, which is the case of the start-up procedure of the engine. The Brayton's model must, therefore, be complemented with the actual design of the turbo-machinery of the engine's stages.

We develop a gas-dynamical model for describing the thermo-fluidic phenomena present in each gas turbine engine's component. For doing so, we apply the fluid's continuity balance for every component of the engine. Especially those with transient mass storage, namely the combustion chamber where compressible effects of the air are relevant and for the coupling of mass fluxes between consecutive components. We also apply the fluid's momentum balance at the core flows in the compressor and turbine stages. Fluid's energy balance accounts for the combustion chamber heat addition and the compression and expansion processes. The mechanical energy equation is also applied to the rotor stages, including the inertial and frictional loses terms of the rotor shaft, to model the mechanical coupling between the moving components. Since our model focus on describing the primary flow-path components and avoids the description of the structural behavior of the solid parts of the engine—which are not essential in the engine's operation. We also neglect the heat transfer between the solid metal components and the gas in the energy balance equation. As reported in previous works such as Ref. [18], some other non-flow-path components affect the ability to maintain the operation conditions and accounting for them can be significant to obtain accurate simulations. This is the case of the combustion model, the fuel system or the external loads related to the propeller. Although we acknowledge the importance of these systems, we exclude the engine-inlet or engine-outlet integration (to reproduce the inlet and outlet flows), the engine-aircraft integration (to reproduce structural analysis), the engine-environment integration (for adverse weather and pollution) and those control systems.

The application of the balances to the engine's components results in a 0-dimensional model representing the operation at transient state. This "aero-thermal" model is usually confined to the normal operating range of the engine but in the present work it is aimed at giving the engine response at off-design operating conditions. A practical approach to numerically solving the 0-dimensional model is to implement it in Matlab-Simulink® software [28], which possess a graphical user interface for building algebraic block diagrams that give solution to the governing differential equations. Hence, we present the global block diagram of the governing equations in Figure 2, where the engine model is represented at a component level. An extended explanation of each engine's component and how we represent it, will be developed through the following paragraphs. Moreover, each component's block diagram will be presented through the detailed explanation of each sub-system. We list inputs, parameters, variables and outputs for every block diagram.

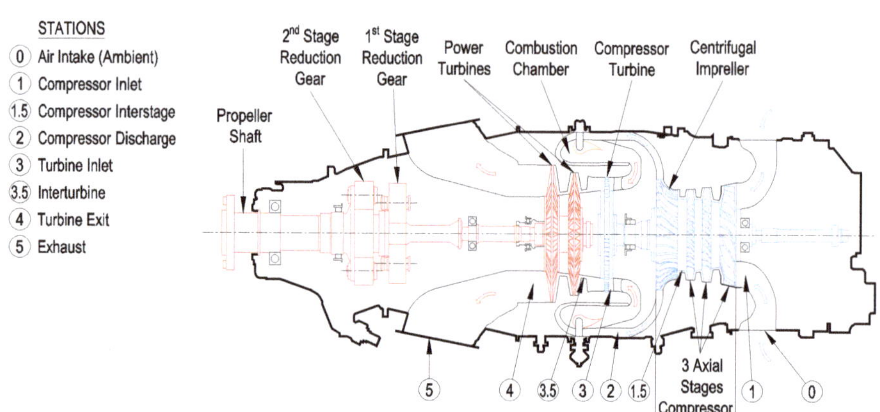

Figure 1. Cross section and stations of the Pratt-Whitney PT6A-65 engine.

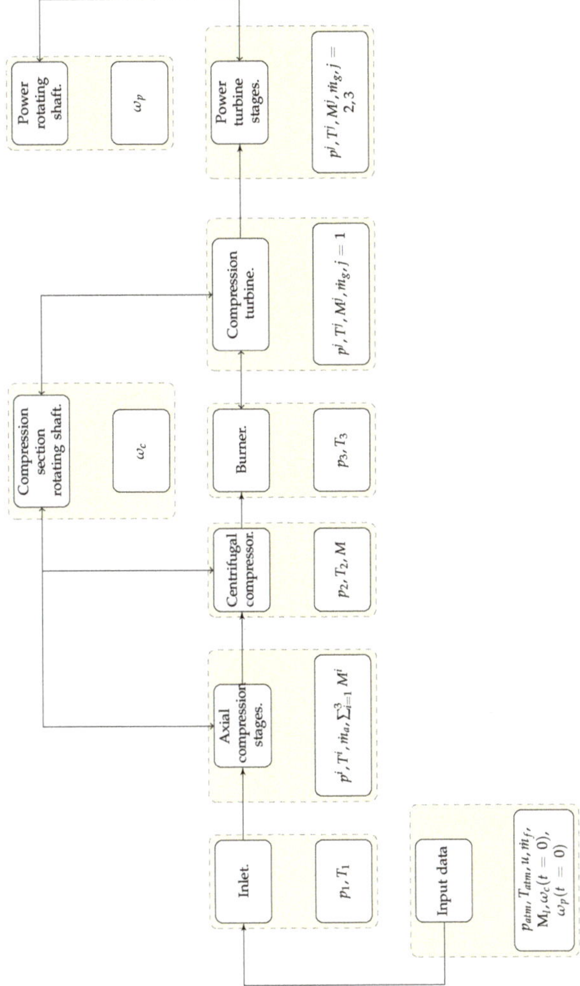

Figure 2. Flow chart of the PT6A-65 engine's model.

2.1. Air Inlet

One of the main characteristics of this gas turbine engine is that it is mounted backward in the nacelle in most aircraft installations. This feature makes that the airflow is directed inside the components of the engine in the same direction as the aircraft's displacement. Another consequence is that the intake is located at the rear part of the engine and therefore, the air passes through the exterior of the aircraft (and the engine itself) before entering to the intake. The PT6A-65 design cares for guiding the intake air to the engine using ducts that avoid facing the exhaust gases: since the typical requirement is to provide laminar and clean air into the compressor—so that, it can operate at maximum efficiency—, the inlet duct changes smoothly from opposing the direction of the airflow to the axially forward direction of the aircraft's speed.

We define the inlet component as to be the intake air conditioner: even though our model is restricted to an on-ground operation, with no rarefying processes of the atmospheric air entering into the engine, we extend the possibility of the operation of the model during in-flight operation conditions. This effect is modeled using the relations for the pressure and temperature of the air which are presented in the block diagram of Figure 3. In the model, p_{atm} and T_{atm} are the atmospheric pressure and temperature, $\text{Ma} = v/c$ is the flight Mach number that relates the aircraft speed v with the speed of sound $c = \sqrt{\gamma p_{atm}/\rho}$. The quotient between specific heats of the air is denoted by γ and η is the polytropic efficiency. We use the subscript 1 to label the thermodynamic variables of air at the inlet. Note that when the engine is supposed to operate in the ground test bench, the model equates the inlet conditions to the atmospheric conditions $p_1 = p_{atm}$ and $T_1 = T_{atm}$.

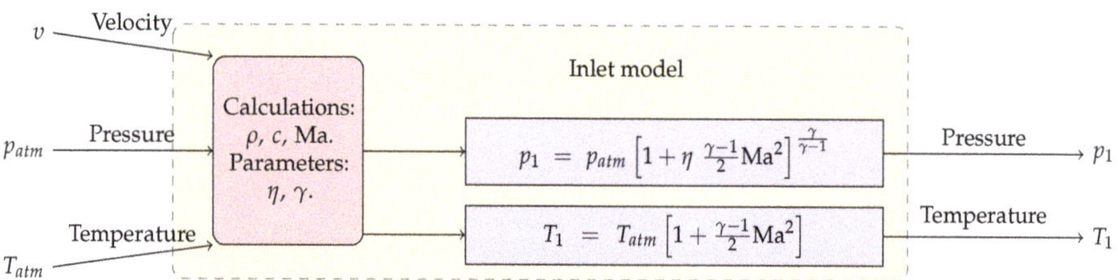

Figure 3. Inlet block.

2.2. Compression Section

The compression section of the PT6A-65 gas turbine engine (used for aviation) consists of three axial stages and a single *centrifugal stage*, each considered to be a rise in the air's pressure: the air flows from the inlet duct to the low-pressure compressor and then to the next two axial-flow stages before passing to the centrifugal stage. The axial stages are composed by rotating blades called *rotors* and static blades called *stators*. Those rotating stages move at around 40,000 Revolutions Per Minute (RPM) increasing the air's pressure more than eight times the inlet's pressure. The last stage of the compressor section is composed of a single centrifugal-flow compressor that accelerates the air outwardly. This centrifugal section has a high-pressure rise but since there can be several losses between centrifugal stages, it is restricted to a single-stage before discharging the airflow. The air stream then leaves the centrifugal compressor section via the diffuser, which is a section of the engine just before the combustion chamber that has the function of preparing the air for the burning area so that it can burn uniformly and continuously. We neglect the diffuser in our representation of the compression section of the engine.

Preliminary work to the modeling of the compressor has to do with the estimation of its geometry. In this sense, we determine the compressor's geometry by extracting most of the information from technical reports, as well as from the engine's manual [29]. Since the axial compression section geometry has been reported in Ref. [30] for a four-axial stages PT6A-65 engine, we transcript the geometrical parameters for the first and last stages but calculate the mean values of the second and third stages of that compressor and set those values as the second axial stage of our PT6A-65 axial compressor model. The geometric parameters that we have processed for each axial stage geometry are presented in Table 1. In the case of the inner and external radius of the rotor blades, we determine those parameters from cross-checking the engine's manual schemes and the measured values from a disassembled engine. The geometry model of the centrifugal compressor is completely defined with the rotor's blade geometry and the inlet's and outlet's cross-sectional dimensions. Where the inner r_1 and outer r_2 radio of the rotor are determined from the engine's technical report, being 92 mm and 117 mm, respectively.

The overall compression stage is defined from the compressor intake (p_1, T_1) to state 2, at the diffuser outlet (p_2, T_2). Therefore, it spans the three axial stages, the centrifugal stage and the diffuser of the PT6A-65 engine. In the present approach, we formulate the mass, momentum and energy balance equations over each sub-stage of the compressor to create a physically-detailed model. Knowledge of the inlet (upwind) conditions of the air, such as temperature and pressure, as well as the rotational speed of the engine's rotating shaft, are required. We also demand the knowledge of the geometry of the compressor at each stage: since each compression section is composed of successive stages of rotating blades (rotors) and stationary guide vanes (stators), we analyze at each rotor-stator stage the transmission of the shaft's mechanical energy into the air's fluid energy and compose the complete compressor performance by adding the multiple successive compression stages.

Table 1. Processed values of aerodynamic and geometric parameters for the axial compressor stages.

Blade	Symbol	1st-Stage Rotor	2nd-Stage Rotor	3th-Stage Rotor
Inlet air angle, deg.	α_l	61.5	59.8	58.0
Exit air angle, deg.	α_t	54.4	48.9	42.0
Inlet metal angle, deg	β_l	56.8	57.85	57.1
Exit metal angle, deg.	β_t	50.5	43.5	36.2
Inner radius, mm.	r_0	72	76	80
Outter radius, mm.	r	100	96	92

For simplicity, we assume that the compressor's blades are thin, rather than having the complete airfoil cross-shape geometry. This supposition is acceptable since in the PT6A-65 engine those are constructed of sheet metal. In the case of the axial compressor, we analyze a single $i-$th stage, where a rotor precedes a stator and consider the same hub radius for the rotor as for the stator. We also make this supposition for the shaft radius at each axial stage. Nevertheless, we consider different cross areas between consecutive compression stages, such that the axial component of velocity can be calculated to conserve mass: in the case of the multistage axial-flow compressor, the blades of each successive stage of the compressor get smaller as the air gets further compressed. In each rotor-stator stage, we consider the cross-section of only one stator and rotor blade as it moves vertically, knowing that the next rotor blade passes shortly thereafter. This is the well-known cascade two-dimensional approximation of turbomachinery analysis and which configures our control volume.

We aim to calculate the variation in the air's enthalpy originated by the rotor's torque. To do this, we apply a simplified analysis of the fluid's dynamics of the air occurring through the blade: the *velocity triangle*. With this approach, no component map representations or error relationships for the guesses of the unknowns are required and therefore, no convergence issues of the numerical solution are found. Instead, direct physical solutions are encountered for each component with the corresponding avoidance of convergence problems. To do this, the axial speed of the airflow v_a^i can be measured at the inlet section, such that the volume flow rate can be calculated in terms of the cross-sectional area. Another possibility is to know the inlet air angle (presented in Table 1), such that the axial velocity is calculated from the velocity triangle. To evaluate the torque on the rotating shaft, we use the angular momentum balance that states that the total momentum in the shaft $\boldsymbol{M}$ is equal to the change between the angular momentum of the flow that crosses the surfaces of the control volume. This change is only related to the tangential velocities of the air given by the velocity triangle analysis.

In our model, we consider irreversible losses in the compression stages and therefore, we include an irreversible mechanical efficiency η that implies a gap between the shaft power and the power delivered to the air. We suppose a quasi-static and polytropic process in which the increment of energy in the air is given by the addition of mechanical power. Indeed, the temperature rise inside the axial compression stage is calculated by applying the energy balance in the control volume. The pressure ratio Π (or net pressure head) induced by the compressor is calculated with the polytropic process relation knowing the temperature variation of the air and the polytropic constant of the gas n. The previous exposition is demonstrated in the block diagrams of Figures 4 and 5.

For the centrifugal compressor, we consider that the circumferential cross-sectional area can be defined by the radius and the width of the blade b. We also suppose that the flow is defined completely in the normal direction $(v_1)_n$ and therefore, the normal velocity at the outlet of the blade can be calculated with the conservation of mass.

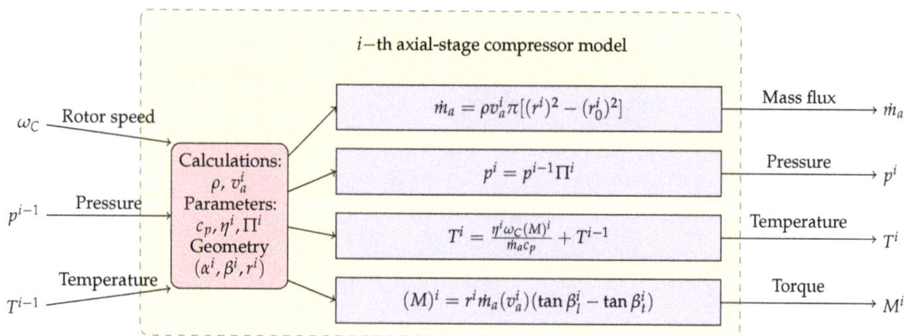

Figure 4. Axial stages compressor block.

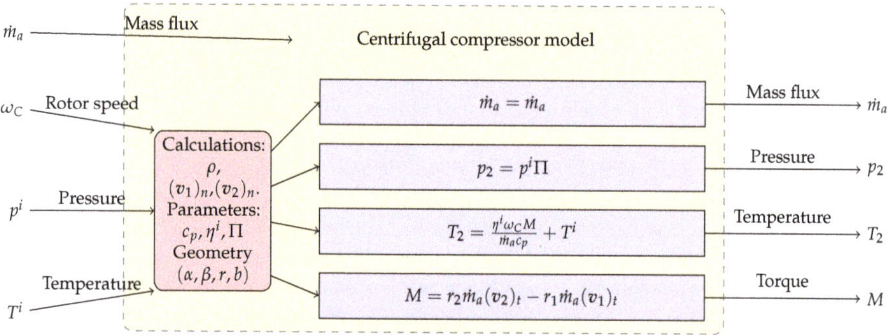

Figure 5. Centrifugal compressor block.

2.3. Burner

The PT6A-65 burner is mainly characterized by the split in the amount of compressed air that is used for maintaining the combustion: only a fraction of the air entering in the burner reacts with the fuel, while most of the compressed air is used for cooling purposes. The geometrical shape of the burner chamber is an annulus. In this sense, the overall volume is hard to be determined and we have approximated its value from measurements of the disassembled engine's burner to be around 0.028 m^3.

Inside the burner, the fuel and the air are separated apart before the flame: combustion with the liquid fuel is performed by the injection of the fuel inside the air stream. Scattering of the fluid into fine droplets leads to the convection and final evaporation of the liquid inside the compressed air-stream. This mixture is a steady non-premixed stream of air-fuel before the full combustion reaction takes place inside the burner. Buoyant mechanisms but mostly forced convection and turbulence mechanisms of inlet air (due to its high pressure and temperature at the exit of the compression section) maintain the combustion process in the flame. The reaction rate and the products of the combustion (exhaust gases) depend on the quality of the air-fuel mixture occurring before the flame.

The combustion quality is automatically controlled by the fuel control system that fixes the amount of injection; this control is set by default for the Jet-A1 fuel. We neglect the fuel control in our present approach since we aim to investigate the response of the engine to different blends with biodiesel. The change in the physical properties of the fuel affects its spraying as it passes through the fuel injectors in the combustion chamber. This has effects on the maximum temperature inside the fuel chamber at the start-up of the engine. In this sense, the starting procedure has to be designed to be rigorous and must be established for each new operating fuel. For all the above, computational simulation of the particular gas turbine engine's performance is proposed as a predictive tool for the engine operation, which allows determining the operation variables when the change in the composition of the fuel occurs. This, with a low cost and without risking the operation of test engines.

Our approach to model the burner's combustion phenomena is simple and follows previous approaches like Ref. [27]. The net power of the gas turbine engine is related to the amount of fuel that is burned inside the burner—the heat added to the air-stream is calculated from the energy balance inside the burning chamber, where LHV is the Low Heating Value of the fuel that represents the amount of energy that is delivered in the combustion process (accounting for the steam boiling in the liquid fuel). This is depicted in the block diagram of Figure 6.

Note that the amount of chemical energy that is transferred to the air (which is later transformed to mechanical energy in the turbine) depends on the type of fuel used and this is completely characterized by the fuel flow rate and its lower heating value. The burner also works as an accumulator of mass, where the temporal change of the thermodynamic conditions of the air inside the burner is related to the amount of fuel $\dot{m}_f$, the incoming airflow rate $\dot{m}_a$, the exhausting rate flow of gasses $\dot{m}_g$ and the volume of the chamber V_B. Note that the fuel flow is the input specification for the fuel burner model, not the Fuel-to-Air Ratio (FAR) or the combustion temperature. The pressure drop across the burner has also been adopted from Ref. [27], with the loss coefficient $C_b \leq 1$, depending on the square mass flow rate.

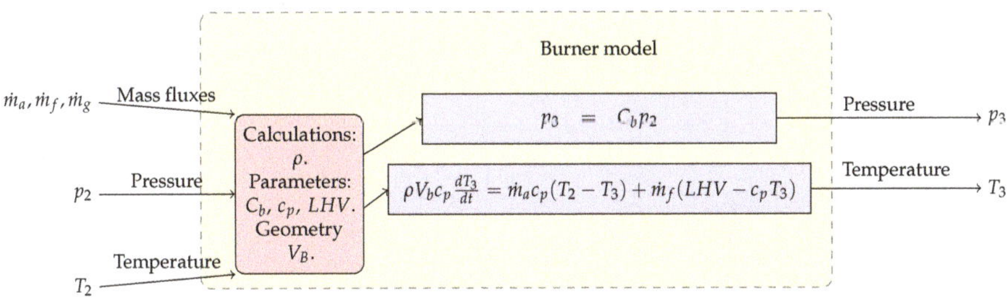

Figure 6. Burner block.

2.4. Turbine Section

The air stream leaves the burner with the addition of heat from the combustion and flows through several turbine stages. The first turbine stage is a single-stage axial turbine that powers the compression section synchronously rotating at 40,000 RPM via the *engine spool* or common shaft. From measurements of the disassembled engine spool, we determine an overall mass moment of inertia of the engine spool of about 0.12 kg m^2.

In the PT6A-65, the hot air flows then into the *power turbines*, which are composed by two axial stages that turn at about $30,000$ RPM and that are connected to the main shaft that drives the propeller (or load). We also determine that the mass moment of inertia of the main shaft is 0.06 kg m^2. The air is discharged next to the exhaust and then to the atmosphere, where the air recovers its original free-stream conditions.

The turbine process is defined from the state 3 at the combustion chamber outlet (p_3, T_3) to state 4, at the engine outlet (p_4, T_4). This means that the expansion ratio of the gas is known for the turbine section. Indeed, knowing the expansion ratio Π^j of the $j-$th stage, one can model each turbine stage as a polytropic expansion process where the pressure and temperature of the air at the discharge can be calculated straightforwardly.

The temperature drop is then used to calculate the retrieved mechanical power inside the turbine. Again, we suppose an irreversible process efficiency between the fluidic and the mechanical power, such that the extracted torque at each axial stage of the turbine differs from the change in the angular momentum of the air inside the turbine. Similarly to the axial flow compressor analysis, an accurate model of the turbine performance can be derived from a detailed computation of the aerodynamics of the flow over the individual blade elements. The distinctive feature of the turbine is that the mass flow rate of gases through the turbine section depends on the expansion work: we use the angular momentum balance to obtain the mass flow of gases through the turbine stage and therefore, the axial term of the air velocity. The turbine relations are presented in the block diagrams of Figures 7 and 8.

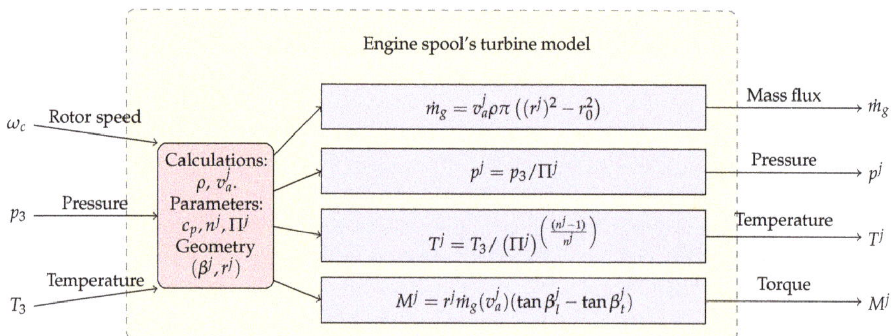

Figure 7. Engine spool's turbine block.

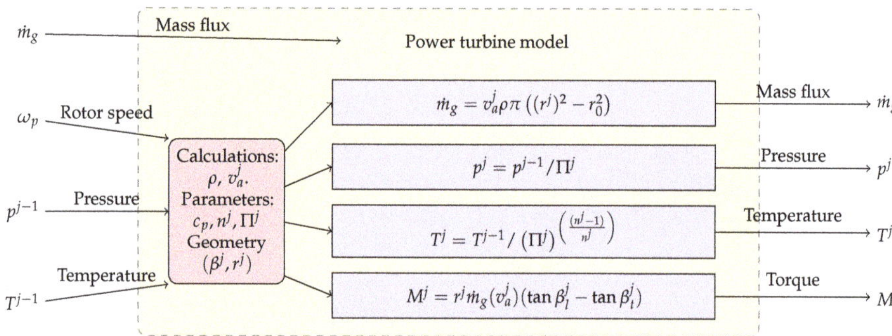

Figure 8. Power turbine block.

2.5. Rotating Shafts

The rotating shafts are modeled by applying the balance of angular momentum. We use the rigid body assumption and apply the momentum balance in the rotational motion, such that the acceleration power of the shaft must equal the balance between turbine power, compression (or load) power and parasitic powers. The angular momentum balance applied to the engine spool, as well as the power rotating shaft are presented in the block diagrams of Figures 9 and 10, respectively. We define I_R to be the mass moment of inertia of the rotating shaft about its rotating axis, $\boldsymbol{M}$ is each one of the torques applied to the shaft and ω is the angular velocity of the shaft which can be calculated in terms of $\omega = N\pi/60$, being N the revolutions per minute. The sum of power losses due to mechanical friction is considered in the torque M_f, which is modeled using a loss-factor b that is a function of the rotational speed of the shaft and its effect acting in the contrary-rotation sense. Following Ref. [27], we apply this mechanical losses model instead of including a mechanical efficiency parameter of the shaft. This is explained since the numerical accuracy of the model can be granted at the idle condition and during the start-up procedure when the output power is negligible.

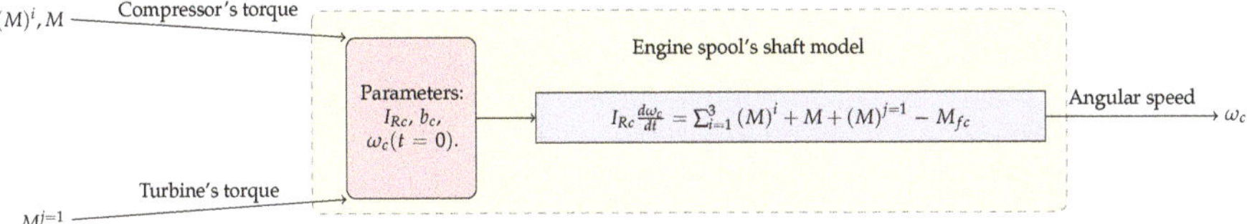

Figure 9. Engine spool's rotor shaft block.

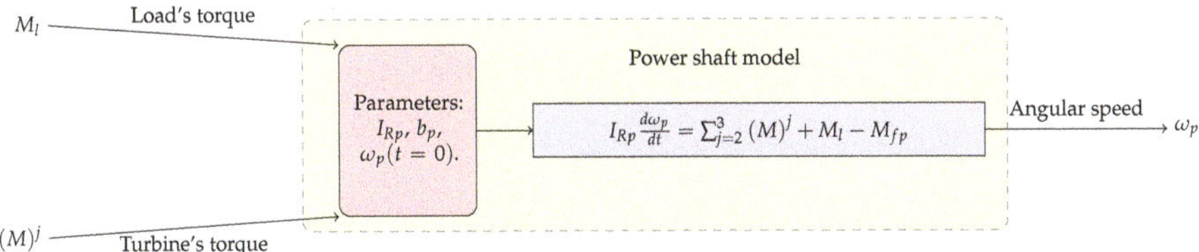

Figure 10. Power shaft block.

2.6. Compatibility Conditions

Besides the physical description of each engine's component, one must close the engine's model coupling the different components to what is referred to the *compatibility conditions*. The overall compatibility conditions become clear from the engine system's block diagram depicted in Figure 2.

The mass compatibility conditions are related to the conservation of the air mass flow rate. The inlet's air mass flow rate $\dot{m}_a$ is calculated by knowing the inlet air angle at the first axial stage, such that the axial velocity can be calculated knowing the rotational speed and the geometric parameters of the rotor blades. It has been explained that $\dot{m}_a$ is conserved among the stages of the compression section. Nevertheless, the mass compatibility condition differs substantially when the air enters the burner: the exhaust gasses mass flow rate $\dot{m}_g$ is not only related to the compressed air flow $\dot{m}_a$ reacting with the mass flow of fuel $\dot{m}_f$ but the exhaust gasses depend on the flow through the turbine section and the engine's exhaust. Here the compatibility condition is related to the mass flow rate resulting from each turbine stage. Some further explanation about this compatibility condition will be taken in the next section.

On the other hand, the angular momentum compatibility condition is the balance of the torques applied for each rotating shaft. Its readily understood that the rotation speed of the axial and centrifugal compressor stages matches with the compressor's turbine via the engine spool. In the same line, the velocity of the power turbines rotating shaft matches the propeller's shaft velocity through the reduction gear.

Finally, the energetic compatibility conditions are associated with the thermodynamic variables of the air at each one of the stages of the engine. It has been readily mentioned that the air enters each consecutive stage with the pressure and temperature conditions that it obtains at the stage immediately before.

3. Numerical Results

In this section, we present the numerical results for several simulation scenarios. As explained before, the numerical solution of the set of algebraic and ordinary differential equations is achieved by Matlab-Simulink®. A fourth-order Runge-Kutta numerical scheme for the temporal integration is used for solving the differential equations. A maximum integration step of 1ms is used to guarantee the stability of the temporal solution. The first scenario is the steady response of the engine which is intended to validate the computational model. Two models are set up, one with the present approach and the other one using GasTurb® [23]. We test both models to ensure the accuracy of the developed model at steady-state operation conditions, and to validate our results with the engine test data and the commercial software simulation. Then, we solve the steady-state operation by using the blends with biodiesel and compare our numerical results with the ones obtained with GasTurb. Finally, we address the transient operation of the engine using the fuel blends, specifically at the start-up of the engine when the maximum temperature can be reached. We suppose in all scenarios an on-ground operation so that the inlet velocity is zero.

3.1. Validation of the Computational Model

We first validate the computational model by considering the standard operation of the PT6A-65 engine reported in the operation manual [29]. Those operation parameters are presented in Table 2, where the International Standard Atmosphere (ISA) conditions are set as the environmental conditions. We evaluate the stationary response of some tracked variables (invariant in time): for the sake of validation, we track the stationary pressure, temperature and airflow at the stations of the engine. In Table 3 we list the experimental results that have been reported in the operation manual for several stations of the engine, and that we use for the sake of comparisons.

Correspondingly with Section 2, the geometrical parameters of the PT6A-65 motor described in that chapter are implemented in our computational model. We fit some remaining geometric parameters so that we obtain the closest numerical results to the ones reported in the operation manual: first, we determine the compression and expansion ratios based on the experimental measurements. The overall compression ratio at the axial stages can be calculated from the data in Table 3 to be 3.14:1 atm, while for the centrifugal compressor the compression ratio is around 2.56:1 atm. In this sense, we fit the centrifugal compressor's angles β_1 and β_2 to 40 and 38 degrees, respectively, such that the centrifugal compressor gives rise to the pressure change. On the other hand, the expansion ratio for the compression and power turbines are calculated to be 1:3.03 atm and 1:2.18 atm, respectively. We suppose an expansion ratio of 1:1.47 atm for each stage of the power turbines section and fit the blade's angles to fulfill the mass flow rate compatibility (conservation) condition between the different turbine stages.

Another quantity that the computational model requires is the n polytropic index for each compression and expansion stage. We calculate that index by solving the logarithmic inverse expression for a polytropic process and accounting for the experimental PT6 engine operation. This is, considering the data in Table 3, one can obtain the polytropic index with the expression

$$n = \left(1 - \left(\frac{\log T_f - \log T_i}{\log p_f - \log p_i}\right)\right)^{-1}.$$

Subscripts i and f of the previous relation denote the initial and final condition of the gas in the polytropic process, respectively. Then, the irreversible efficiency η of each compression and expansion stage can be computed from $\eta = \left(\frac{\gamma-1}{\gamma}\right) / \left(\frac{n-1}{n}\right)$, where the $\gamma = 1.4$ value corresponds to the air (not to be confused with the overall engine's efficiency). The gathered polytropic indexes and irreversible efficiencies for the compression and expansion sections of the PT6A-65 engine are presented in Table 4. Those values are set in the energy balances and polytropic processes of the compressor and turbine stages.

Table 5 presents the fitted aerodynamic and geometric parameters for all the turbine stages of the PT6A-65 engine that required to be processed. These parameters are determined following the previously exposed ideas, from cross-checking the engine's manual schemes and the measured values from a disassembled engine but mostly from the fulfillment of the mass conservation requirement. The rotor blade airfoils, which are metal profiles followed and preceded by stator vanes, are completely defined by these geometric parameters. Finally, the parasitic power that is lost due to friction can be modeled by setting the bearing coefficients b_c and b_p to 0.04 kg·m^2/s and 0.05 kg·m^2/s, respectively, for the engine spool and power rotor shaft. We also model the pressure loss coefficient in the burner to be $C_b = 0.95$. These values arise from the numerical experiments that we performed guaranteeing the best approximation of the computational model to the referenced values of the engine's steady operation.

The simulated steady engine response is presented in Table 6. In those results, we have assumed a constant flow of Jet-A1 fuel, with a calorific power of 42.8 MJ/kg and a constant propeller's load torque of 2684.51 N·m. We confirm a consistent physical behavior of the thermodynamic variables with the experimental reported ones. Distinctively, the observed pressure at the discharge of the axial compressor stages is higher than the reported one. This inaccuracy is countered by the centrifugal compression section, where the blade's geometrical parameters are fitted to give accurate results of the overall compression ratio. In this regard, fitting the compressor's parameters affects negatively the accuracy of the temperature at the discharge of the compression stage but the energy balance inside the combustion chamber counteracts this effect and matches the stipulated temperature in the manual, with an error of only 1.57%.

Temperature and pressure variables at the expansion stages correspond well to the experimental counterparts, mainly due to the possibility of fixing the expansion relation and the geometrical parameters of the turbines. Given the previous exposition, we believe that the error is restricted to a low range such that it validates the usage of the proposed model for predicting the engine response when new operating conditions are evaluated.

As explained before, we also implement the PT6 engine in the GasTurb software to validate our results: we simulate the stationary operation of the engine using the Jet-A1 fuel and the same operating conditions that have been listed in Table 2. Since the GasTurb software has three different simulation levels: the basic thermodynamic approach, an engine's performance mode and a more advanced engine's performance simulation that considers the specific geometry of the gas turbine engine, we choose the engine's performance simulation. We select the default two-spools turboprop template that includes a booster and a compressor, the combustion chamber and the two expansion stations but modify the template by including a high-pressure turbine in the engine's spool and a low-pressure turbine in the power spool. The inputs for simulation in GasTurb are also the polytropic efficiencies of compressors and turbines, the atmospheric condition (temperature and pressure), the drop pressure in the combustion chamber, the low heat value of the fuel, the maximum temperature in the combustion chamber and the pressure ratios in the compressor and turbine stations. The temperature and pressure results along the stations of the PT6A-65 engine, together with the calculated relative error against the experimental measurements are presented in Table 6. The comparisons between the developed model results and GasTurb demonstrate that although there are some errors for certain stages in each model, the results from the present approach do not differ in accuracy than those given by GasTurb. These results guarantee an acceptable resolution of the present approach and give the certainty to apply it in the PT6A engine evaluation when operating at off-design conditions.

Table 2. On-ground steady operation conditions using Jet-A1 fuel. Extracted from Ref. [29].

Standard Conditions	Value
Atmospheric temperature	288 K
Atmospheric pressure	101,352.9 Pa
LHV of Jet-A1 fuel	42.8 MJ/kg
Jet-A1 fuel mass flow	0.062 kg/s
Propeller's load (at propeller's shaft)	2684.51 N.m

Table 3. Temperatures and pressures for PT6A-65 engine at 850 shp and ISA standard conditions. Extracted from Ref. [29].

Station	Location	Temperature (K)	Pressure (Pa)
0	Ambient	288	101,352.9
1	Compressor Inlet	288.2	102,042.4
1.5	Compressor Interstage	415.4	307,506.2
2	Compressor Discharge	610.4	787,381.28
3	Turbine Inlet	1212.1	770,144.39
3.5	Interturbine	967.1	246,142.8
4	Turbine Exit	811.5	113,074
5	Exhaust	811.5	106,179.3

Table 4. Polytropic index and efficiency gathered values for the compression and expansion sections of the PT6A-65 engine.

Description	Stages	Polytropic Index n	Irreversible Efficiency η
Axial Compressor	1–1.5	1.496	0.862
Centrifugal compressor	2.5–3	1.693	0.698
Hot turbine expansion	3–3.5	1.247	1.442
Cold turbine expansion	3.5–4	1.291	1.267

Table 5. Fitted aerodynamic and geometric parameters of the turbine stages.

Blade	1st-Stage Rotor	2nd-Stage Rotor	3th-Stage Rotor
Inlet metal angle, deg	0	0	0
Exit metal angle, deg.	80	43	43
Inner radius, mm.	92	90	88
Outter radius, mm.	117	125	142

Table 6. Present and GasTurb simulation results. Stationary temperatures and pressures at different stations of the PT6A-65 engine. The relative error is calculated against the reported results in Table 3.

Station	Present Simulation				GasTurb Simulation			
	T(K)	Error(%)	p(pa)	Error(%)	T(K)	Error(%)	p(pa)	Error(%)
1.5	433.63	4.39	423,635.89	37.76	420.11	1.12	318,248	3.37
2	510.44	16.38	749,710.57	4.78	611.03	0.10	814,715	3.35
3	1289.15	6.35	727,219.25	5.57	1212.1	0	773,979	0.48
3.5	925.3	4.11	240,006.35	2.49	863.07	12.05	137,405	44.17
4	759.69	6.38	111,067.77	1.77	829.8	2.20	122,816	7.93

3.2. Stationary Operation of the PT6A-65 Using Fuel Blends

Once the present computational model has been validated, the following simulation scenarios are considered: we evaluate the stationary operation of the PT6A-65 engine at both 100% of the fuel mass flow and 60% of the fuel flow, meaning full throttle and idle operation, respectively. Since we aim to predict the engine's response to the usage of new hypothetical fuels, we vary the parameters related to the fuel's modeling. In this sense, the standard Jet-A1 fuel—kerosene—is mainly composed of $n-$heptane and isooctane, which are hydrocarbons that possess between 8 and 16 carbon atoms per molecule, giving a LHV of around 42.8 MJ/kg (measured in experimental tests [31]). On the other hand, the chemical composition of a biodiesel sample has been measured in Ref. [32] giving an approximated LHV value of 36.29 MJ/kg. Hence, the pure biodiesel retains a smaller amount of energy than conventional Jet-A1 fuel.

We simulate the engine operation with hypothetical blends of Jet-A1 with biodiesel. These fuel blends are abbreviated as KB#, where "K" represents the Jet-A1 as the primary fuel, "B" represents the biodiesel as the secondary fuel and # is the mass fraction of biodiesel in the blend expressed in percentage. We establish a discrete range of mass concentrations of biodiesel in Jet-A1 which are 3 in total: KB10, KB20 and KB30 and calculate their LHV using the percent by mass of the mixture. The KB0 and KB100 represent the pure Jet-A1 and pure biodiesel fuels, respectively. In the successive, we plot the simulation results for the operation with different blends of fuel, so that they are easily comparable. Table 7 shows the LHV for each blend and the notation to be used in the plots.

Table 7. Biodiesel blends.

Blend	LHV (MJ/kg)	Notation in Plots
KB0	42.8	○
KB10	42.14	⋆
KB20	41.49	△
KB30	40.84	*
KB100	36.29	□

Figure 11 shows the air pressure through the engine stages. We observe a pressure rise at the compression section, as well as an expansion process at the turbine stages. It is evident for both throttle operations that the maximum pressure is located at the centrifugal compressor's discharge (stage 2) and that there is a slight loss in this pressure at the burner (stage 3). In the full-throttle case (right side of the figure), it can be observed that the pressure gap between the Jet-A1 fuel and the pure biodiesel fuel is considerable, with an observable maximum pressure of 750 kPa and a lower pressure of 610 kPa. In the case of the idle operation, there is a difference of 60 kPa between those fuels, with a maximum pressure of 460 kPa and a minimum pressure of approximately 400 kPa. The fuel blends give results that lay below some 5% of the Jet-A1 pressure range. In any case, the maximum values of pressure are related to the use of the Jet-A1 fuel. This can be explained since it delivers the highest amount of power, which in turn is extracted by the turbine stage and transferred to the compression section via the engine spool.

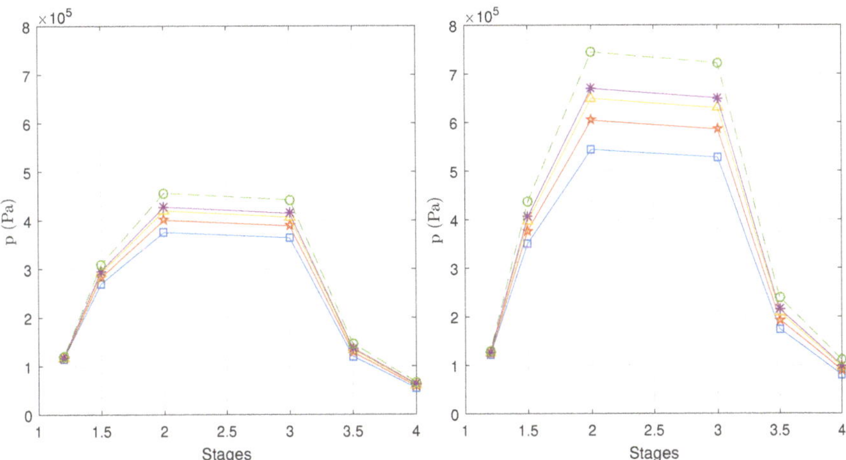

Figure 11. Pressure distribution along the engine stages. Results for the stationary engine operation with the 60% (**left**) and 100% (**right**) throttles.

Figure 12 displays the air temperature through the engine stages for the two different throttles. The temperature results agree well with those reported in the manual: the maximum air temperature is observed at the burner discharge (stage 3). In any scenario, the fuel that provides the highest temperature to the engine is the Jet-A1 fuel. For the full-throttle operation there is a 200 K temperature difference in the burner between the Jet-A1 fuel and the pure biodiesel fuel: the maximum temperature is given for the Jet-A1 fuel, reaching 1290 K, while the minimum temperature of 1104 K is given by the biodiesel operation. The highest drop in temperature occurs at the compression turbine, while the power turbines contribute lesser to the temperature reduction. It is also observed that the temperature difference arising between the fuels at the burner is kept constant at the turbines discharge (stage 4). No important temperature variation is observed for the biodiesel blends. Regardless of the fuel type, the temperature never exceeds a value of 1050 K for the idle operation.

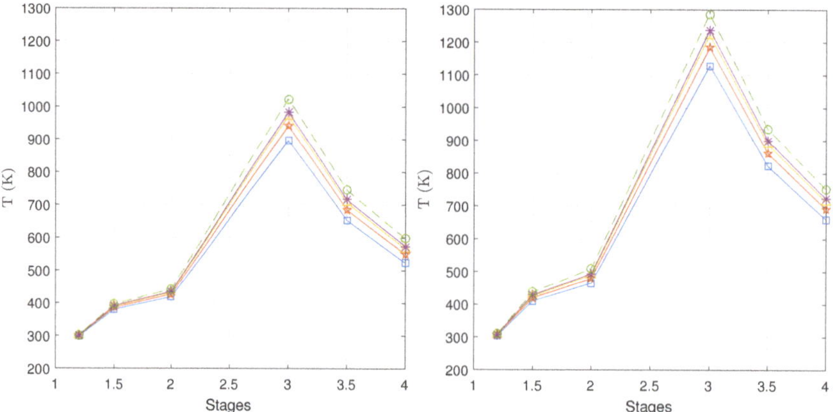

Figure 12. Temperature distribution along the engine stages. Results for the stationary engine operation with the 60% (**left**) and 100% (**right**) throttles.

We also use GasTurb software to evaluate the engine performance running with biodiesel blends. Hence, we display in Figure 13 the pressure and temperature results obtained with the GasTurb model for the full-throttle operation. We can observe that the GasTurb results for the different mixtures do not vary significantly. Although we have imposed modified LHV values for each fuel, the GasTurb model seems not to be sensitive to these variations. On the contrary, in our results, we can appreciate the variations in pressure and temperature that were commented previously.

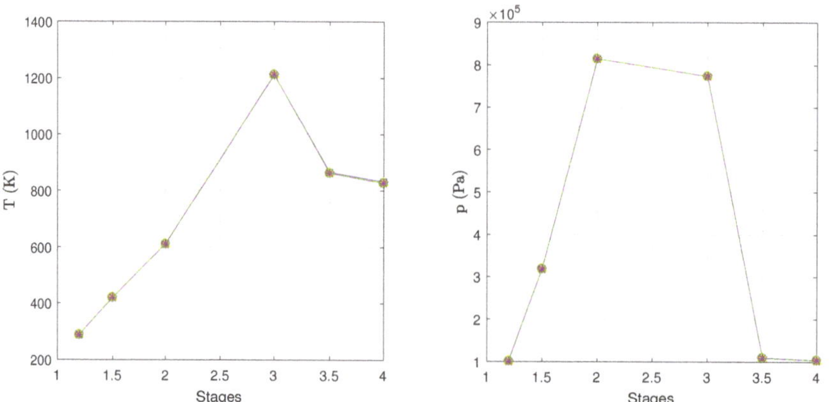

Figure 13. GasTurb results . Temperature (**left**) and pressure (**right**) results for the stationary engine operation with the 100% throttle.

3.3. Transient Operation of the PT6A-65 Motor Using Fuel Blends

Finally, we evaluate the start-up procedure of the engine with the fuel types that have been tested in previous scenarios. The main goal is to identify problematic conditions during the engine start-up, which is the transient procedure that can actually affect the engine's integrity. The model is intended to run at zero speed and sea level conditions, since the off-design operation of the engine are evaluated on a test bench. Being a two-spool turboprop, the PT6A engine has been designed to be electrically started. The starter motor is represented in the engine simulation as a constant rotational speed and starter torque. Hence, at the start up process, a simplified model with the imposition of the rotational speed at the compression shaft and a constant torque load is implemented. The simulation is also intended to run from idle to maximum power but no shut down simulation is proposed since the objective is focused on the combustion chamber's limits. We configure the start-up procedure such that the compressor's shaft starts accelerating from its steady state condition to $N_c(t = 0) = 12{,}000$ RPM, which is the rotational speed that is provided by the starter. From this point, there is a positive increment of the air pressure (given by the rotation of the compression system) until the desired mass flow of air into the burner is granted. During the initial compression operation, no fuel mass flow is injected into the burner. We consider that at a later instant $(t = 10)$ s, when the compressed air into the burner stabilizes, the fuel is injected into the burner and ignited. The mass fuel flow is then gradually increased until the desired throttle is reached at $(t = t_f)$. All the start-up procedure is considered to undergo with a constant torque at the propeller's shaft $M_l = 2684.51$ N·m.

We aim to evaluate the start-up procedure of the PT6A-65 engine with both 100% and 60% of throttle. The results for the operation with fuel blends are displayed similarly as for the stationary operation. We present comparisons of some important variables, such as the air pressure at the compressor discharge and the air temperature inside the burner. It is noticeable that the start-up procedure converges to the steady-state operation.

Figure 14 shows the transient pressure results at the compression stage for the two different throttles. It can be observed that the pressure in the compressor's discharge undergoes an initial equilibrium when the engine spool is started, reaching a compression ratio below of 1.5:1 atm. At $(t = 10)$ s, when the fuel is ignited, a sudden increment of the compression ratio of around 2:1 atm is noticed for all fuels and throttles. In the case of the 60% throttle, the pressure stabilizes from this instantaneous peak but a pressure increasing delay for the full throttle case is noticeable, reaching lately the reported 8:1 atm compression ratio. There is not a significant pressure fluctuation related to the fuel blends: it can only be appreciated a moderate increment for the full throttle and the biodiesel fuel than for the blends and Jet-A1 fuel.

The transient temperature inside the burner is presented in Figure 15 for the two different throttles. We observe a slight increment in the temperature at the initial compression operation. Then, the fuel intake generates a temperature peak in the combustion chamber, which in all scenarios is the maximum temperature that is reached during the complete operation of the engine. Since the initial fuel flow is the same for both throttle scenarios, the maximum temperature in the engine does not vary. Instead, it only depends on the fuel blend, where the maximum temperature of 1300 K inside the burner is obtained with the Jet-A1 fuel, and the minimum of around 1100 K, approximately, is obtained with the pure biodiesel. This is explained by the relatively smaller LHV of the biodiesel, which affects the gas turbine combustion. After this critical instant, the temperature stabilizes at around 1000 K for the idle operation, with fluctuation slightly greater than 100 K between the JetA1 and the biodiesel fuel. On the other hand, the temperature in the full-throttle scenario increases gradually until the steady-state of 1290 K is reached at about 40 s.

The first remark up to this point is that temperature controls have not been implemented in the present model. Instead, we want to establish the hypothetical temperature that is reached during the ignition procedure. Using the present computer engine model we can compare the start-up operating temperature against the reference given by the engine test data, which is the limit for structural integrity reasons and shutdown limit. In this sense, the standard 1300 K temperature limit [29] is not reached by any off-design operational scenario. The tests of the engine operation with biodiesel blends show that the maximum temperature is lesser than 1220 K, which is acceptable for structural reasons of the engine. However, in the real engine operation, the fuel is introduced by the fuel control system. Since the accumulation of unburnt fuel can result in a flammable mixture that can potentially lead to an explosion in the engine, the real combustion with biodiesel blends is controlled by this system. We acknowledge the limitation of not including the fuel system but consider it as future work to expand the present model.

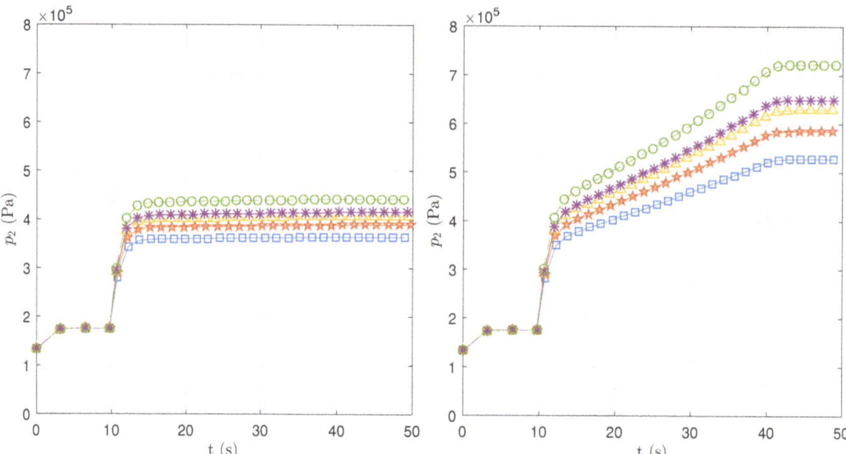

Figure 14. Transient pressure at the compressor's discharge. Results for the start-up operation with the 60% (**left**) and 100% (**right**) throttles.

Another remark is that the computational simulations given by Matlab-Simulink agree with the actual physical behavior of the engine, guaranteeing the numerical convergence to the solution at every time step of the transient simulation.

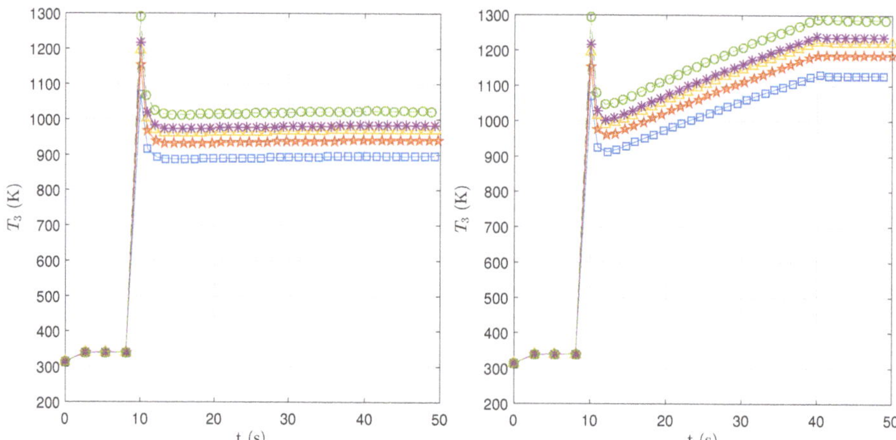

Figure 15. Transient temperature at the burner. Results for the start-up operation with the 60% (**left**) and 100% (**right**) throttles.

4. Conclusions

In this article, we have developed a functional PT6A-65 engine computer model, which can be used for purposes of testing off-design operating conditions with blends of biodiesel fuel. For this, we have simplified the PT6A-65 engine in a process that extracted the essential components of the engine and eliminated the auxiliary ones. The mass, linear momentum, angular momentum and energy balances have been applied into these components, by which together with the compatibility conditions compose a 0-dimensional aero-thermal model of the engine. We have implemented the numerical solution of the model in the Matlab-Simulink® software and different simulation scenarios have granted the predictive capacity of our computational model concerning the systemic and transient response of the engine. Thus, protecting the actual PT6A-65 plant and incurring in considerable computational advantages like accuracy, economy, expedition and applicability.

Indeed, the achievements of the computer model approach are aligned with our expectations since they have given accurate descriptions of the engine's performance, not only in the sense that the model has been tuned with the real PT6 operation but also because those are governed by physical expressions. We have obtained the expected performance in terms of desired computational cost. This is one big strength of the present approach, to be able to obtain transitory solutions for the overall engine's operation in a personal computer. Certainly, this model is a fair trade-off of accuracy/costs since complex CFD models cannot represent the engine's response in a systematic approach, those can only handle the transient response of very few elements and very limited time ranges.

One general remark is that the dynamical system's approach (given in algebraic blocks) may be applied to other gas turbine engines: this by modifying the turboprop arrangement with a new connection between the outputs and inputs of the various blocks and without spending too much effort in developing a new computational model. Another is that the present computer model is complimentary to widely used commercial software like GasTurb®. Since the present results and the ones obtained with such software have been compared, we have granted that steady descriptions of the PT6 engine given by our model are at least as accurate and reliable than those by GasTurb®; The relative error between the present results and those reported in the manual is below 10% for most of the stages. But also, that the transient analysis gives a clear advantage to our computational approach.

Concerning the evaluation of the engine's response to different blends of Biodiesel (or other novel types of fuels), we have obtained the desired predictions: such as the maximum burner's temperature at the start-up process or the pressure distribution through the engine stages. Regarding the structural integrity of the engine, this approach can give accurate predictions of the temperature in the combustion chamber and evaluate if it remains in the appropriate operating range given by the operation manual (overhaul) such that the operation with the fuel blend does not incur in structural damage of the combustion chamber. The simulation results allow us to conclude that the use of blends of biodiesel that are less energetic than the conventional Jet-A1 fuel would not generate a burner overheating since the standard temperature of 1220 K inside the chamber is never exceeded. It is also clear that fuel blends would not generate a pressure excess in the compressor's discharge and, consequently, these can not generate internal damage to the motor structure. Those results argue that the proposed methodology can be used in future gas turbine engine's operation with other fuel blends and the engine's evaluation in off-design operating conditions.

As future work, kinetic combustion models can be coupled to the present model to determine the added heat and the production of gases. Hence, a natural sequel of the present article is to implement engine control: torque control, pressure control, the fuel system and the combustion control. Even though we recognize that with the present approach we can not calculate the thermal efficiency of the computer model engine since the relation between the engine performance (fuel consumption) against the load is not well represented, we expect to develop a combustion model or a combustion control in our computer model that gives accurate responses of the engine's efficiency regarding each fuel consumption. We believe that such a work deserves an article on its own and it is not the motivation of the present study. In this sense, reported higher efficiencies for gas turbines operating with fuel blends in References [25,26], together with our simulation results, drive to continue working on these topics. Also, non-linear brake models stand as other possible development towards the depiction of the real loading conditions of the engine.

Author Contributions: Conceptualization, C.B.-R. and J.S.S.-C.; methodology, C.B.-R.; software, C.B.-R., A.G.R.-M. and D.C.; validation, C.B.-R., A.G.R.-M. and D.C.; formal analysis, C.B.-R.; investigation, C.B.-R.; writing—original draft preparation, C.B.-R.; writing—review and editing, C.B.-R.; visualization, C.B.-R. and D.C.; project administration, J.B.

Funding: This research was funded by Departamento Administrativo de Ciencia, Tecnología e Innovación through the "Uso de Bioqueroseno como Combustible en Aeronaves de la FAC" project.

Acknowledgments: The authors thankfully acknowledge the resources, technical expertise, and assistance provided by the Colombian Air Force (FAC). This work was supported by Universidad ECCI.

Conflicts of Interest: The authors declare no conflict of interest.

Abbreviations

The following nomenclature is used in this manuscript:

Nomenclature		**Greek letters**	
c	Speed of sound	α	Air angle
C_b	Pressure loss coefficient	β	Blade angle
c_p	Specific heat capacity at constant pressure	γ	Quotient between specific heats of the air
I_r	Mass moment of inertia	η	Isentropic efficiency
i, j	Axial stages counters	η_b	Brayton's cycle efficiency
LHV	Low Heating Value	η_r	Brake efficiency
$\boldsymbol{M}$	Torque	Π	Pressure ratio
$m, \dot{m}$	Mass, mass flow	ρ	Density
Ma	Mach's number	ω	Angular velocity
n	Polytropic index	**Subscript**	
N	Number of revolutions per minute	a	Air
p	Pressure	atm	Atmosphere
R	Constant of gases	b	Burner
r	Radius	c	Compressor
T	Temperarture	f	Fuel
t	Time	g	Gas
V	Volume	l	Leading edge
v	Velocity	p	Propeller
		t	Trailing edge

References

1. Gatto, E.; Li, Y.; Pilidis, P. Gas turbine off-design performance adaptation using a genetic algorithm. In Proceedings of the ASME Turbo Expo 2006: Power for Land, Sea, and Air, Barcelona, Spain, 8–11 May 2006; American Society of Mechanical Engineers: New York, NY, USA, 2006; pp. 551–560.
2. Li, Y.; Ghafir, M.; Wang, L.; Singh, R.; Huang, K.; Feng, X. Nonlinear multiple points gas turbine off-design performance adaptation using a genetic algorithm. *J. Eng. Gas Turbines Power* **2011**, *133*, 071701. [CrossRef]
3. Li, Y.; Ghafir, M.; Wang, L.; Singh, R.; Huang, K.; Feng, X.; Zhang, W. Improved multiple point nonlinear genetic algorithm based performance adaptation using least square method. *J. Eng. Gas Turbines Power* **2012**, *134*, 031701. [CrossRef]
4. Jin, H.; Frassoldati, A.; Wang, Y.; Zhang, X.; Zeng, M.; Li, Y.; Qi, F.; Cuoci, A.; Faravelli, T. Kinetic modeling study of benzene and PAH formation in laminar methane flames. *Combust. Flame* **2015**, *162*, 1692–1711. [CrossRef]
5. Kong, C.; Ki, J.; Koh, K. Steady-State and Transient Performance Simulation of a Turboshaft Engine with Free Power Turbine. In Proceedings of the ASME 1999 International Gas Turbine and Aeroengine Congress and Exhibition, Citeseer, Indianapolis, Indiana, 7–10 June 1999; p. V002T04A016.

6. Kong, C.; Roh, H. Steady-state Performance Simulation of PT6A-62 Turboprop Engine Using SIMULINK®. *Int. J. Turbo Jet-Engines* **2003**, *20*, 183–194. [CrossRef]
7. Alexiou, A.; Mathioudakis, K. Development of gas turbine performance models using a generic simulation tool. In Proceedings of the ASME Turbo Expo 2005: Power for Land, Sea, and Air, Reno-Tahoe, Nevada, USA, 6–9 June 2005; American Society of Mechanical Engineers: New York, NY, USA, 2005; pp. 185–194.
8. Cuenot, B. Chapter Four—Gas Turbines and Engine Simulations. In *Thermochemical Process Engineering*; Academic Press: Cambridge, MA, USA, 2016; Volume 49, pp. 273–385. [CrossRef]
9. Singh, R.; Maity, A.; Nataraj, P. Modeling, Simulation and Validation of Mini SR-30 Gas Turbine Engine. *IFAC-PapersOnLine* **2018**, *51*, 554 – 559. [CrossRef]
10. Mcguirk, J.J.; Palma, J. The flow inside a model gas turbine combustor: Calculations. *J. Eng. Gas Turbines Power* **1993**, *115*, 594–602. [CrossRef]
11. Sirignano, W.; Liu, F. Performance increases for gas-turbine engines through combustion inside the turbine. *J. Propuls. Power* **1999**, *15*, 111–118. [CrossRef]
12. Liu, F.; Sirignano, W. Turbojet and turbofan engine performance increases through turbine burners. *J. Propuls. Power* **2001**, *17*, 695–705. [CrossRef]
13. Menon, S. Simulation of Combustion Dynamics in Gas Turbine Engines. In *Parallel Computational Fluid Dynamics 2002*; North-Holland: Amsterdam, The Netherlands, 2003; pp. 33–42. [CrossRef]
14. Smirnov, N.; Nikitin, V.; Stamov, L.; Mikhalchenko, E.; Tyurenkova, V. Rotating detonation in a ramjet engine three-dimensional modeling. *Aerosp. Sci. Technol.* **2018**, *81*, 213–224. [CrossRef]
15. Ghoreyshi, S.; Schobeiri, M. Numerical simulation of the multistage ultra-high efficiency gas turbine engine, UHEGT. In Proceedings of the ASME Turbo Expo 2017: Turbomachinery Technical Conference and Exposition, Charlotte, NC, USA, 26–30 June 2017; American Society of Mechanical Engineers: New York City, NY, USA, 2017; p. V003T06A034.
16. Gaudet, S.R. Development of a Dynamic Modeling and Control System Design Methodology for Gas Turbines; Ph.D. Thesis, Carleton University, Ottawa, ON, Canada , 2008.
17. Panov, V.; Smith, M. Numerical Simulation of Partial Flame Failure in Gas Turbine Engine. In *ASME Turbo Expo 2007: Power for Land, Sea, and Air*; American Society of Mechanical Engineers: New York City, NY, USA, 2007; pp. 523–529.
18. Wang, C.; Li, Y.G.; Yang, B.Y. Transient performance simulation of aircraft engine integrated with fuel and control systems. *Appl. Therm. Eng.* **2017**, *114*, 1029–1037. [CrossRef]
19. Sellers, J.F.; Daniele, C.J. *DYNGEN: A Program for Calculating Steady-State and tRansient Performance of Turbojet and Turbofan Engines*; National Aeronautics and Space Administration: Washington, DC, USA, 1975.
20. Visser, W.P.; Broomhead, M.J.; van der Vorst, J. TERTS: A generic real-time gas turbine simulation environment. In *ASME Turbo Expo 2001: Power for Land, Sea, and Air*; American Society of Mechanical Engineers Digital Collection; American Society of Mechanical Engineers: New York City, NY, USA, 2002.
21. Visser, W.P.; Broomhead, M.J. GSP, a generic object-oriented gas turbine simulation environment. In *ASME Turbo Expo 2000: Power for Land, Sea, and Air*; American Society of Mechanical Engineers Digital Collection; American Society of Mechanical Engineers: New York City, NY, USA, 2000.
22. Panov, V. Gasturbolib: Simulink library for gas turbine engine modelling. In *ASME Turbo Expo 2009: Power for Land, Sea, and Air*; American Society of Mechanical Engineers Digital Collection; American Society of Mechanical Engineers: New York City, NY, USA, 2009; pp. 555–565.
23. Kurzke, J. GasTurb 9—A Program to Calculate Design and Off-Design Performance of Gas Turbines. 2001. Available online: http://www.gasturb.de (accessed on 1 May 2019).
24. Idebrant, A.; Näs, L. Gas turbine applications using thermofluid. In Proceedings of the 3rd International Modelica Conference, Linköping, Sweden, 3–4 November 2003.
25. Nascimento, M.; Lora, E.; Corrêa, P.; Andrade, R.; Rendon, M.; Venturini, O.; Ramirez, G. Biodiesel fuel in diesel micro-turbine engines: Modelling and experimental evaluation. *Energy* **2008**, *33*, 233–240. [CrossRef]
26. Azami, M.; Savill, M. Comparative study of alternative biofuels on aircraft engine performance. *Proc. Inst. Mech. Eng. Part J. Aerosp. Eng.* **2017**, *231*, 1509–1521. [CrossRef]
27. Camporeale, S.; Fortunato, B.; Mastrovito, M. A modular code for real time dynamic simulation of gas turbines in simulink. *J. Eng. Gas Turbines Power* **2006**, *128*, 506–517. [CrossRef]
28. MATLAB Simulink Toolbox. Release 2016a. Available online: www.mathworks.com (accessed on 1 May 2019).

29. Whitney, P. *PT6 TRAINING MANUAL*; Pratt & Whitney Canada Corp.: Longueuil, QC, Canada, 2001.
30. Yoshinaka, T.; Thue, K. A Cost-Effective Performance Development of the PT6A-65 Turboprop Compressor. In Proceedings of the ASME 1985 Beijing International Gas Turbine Symposium and Exposition, Citeseer, Beijing, China, 1–7 September 1985; p. V001T02A015.
31. Chong, C.T.; Hochgreb, S. Spray flame structure of rapeseed biodiesel and Jet-A1 fuel. *Fuel* **2014**, *115*, 551–558. [CrossRef]
32. Llamas-Lois, A. Biodiesel and Biokerosenes: Production, Characterization, Soot & Pah Emissions. Ph.D. Thesis, ETSI_Energia, Madrid, Spain , 2015.

Article

Experimental Methodology and Facility for the J69-Engine Performance and Emissions Evaluation Using Jet A1 and Biodiesel Blends

Gabriel Talero [1], Camilo Bayona-Roa [1,*], Giovanny Muñoz [1], Miguel Galindo [1], Vladimir Silva [1], Juan Pava [2] and Mauricio Lopez [3]

1 Universidad ECCI, Cra. 19 No. 49-20, Bogotá 111311, Colombia; gtaleror@ecci.edu.co (G.T.); giovannya.munoze@ecci.edu.co (G.M.); josem.galindoc@ecci.edu.co (M.G.); vsilval@ecci.edu.co (V.S.)
2 Escuela de Postgrados de la Fuerza Aérea Colombiana, Cra. 57 No. 43-28—CAN Puerta 8, Bogotá 111321, Colombia; juan.pava@epfac.edu.co
3 Fuerza Aérea Colombiana, Cra. 57 No. 43-28—CAN Puerta 8, Bogotá 111321, Colombia; mauricio.lopezg@fac.mil.co
* Correspondence: cbayonar@ecci.edu.co

Received: 5 November 2019; Accepted: 25 November 2019; Published: 28 November 2019

Abstract: Aeronautic transport is a leading energy consumer that strongly contributes to greenhouse gas emissions due to a significant dependency on fossil fuels. Biodiesel, a substitution of conventional fuels, is considered as an alternative fuel for aircrafts and power generation turbine engines. Unfortunately, experimentation has been mostly limited to small scale turbines, and technical challenges remain open regarding operational safety. The current study presents the facility, the instrumentation, and the measured results of experimental tests in a 640 kW full-scale J69-T-25A turbojet engine, operating with blends of Jet A1 and oil palm biodiesel with volume contents from 0% to 10% at different load regimes. Findings are related to the fuel injection system, the engine thrust, and the emissions. The thrust force and the exhaust gas temperature do not expose a significant variation in all the operation regimes with the utilization of up to 10% volume content of biodiesel. A maximum increase of 36% in fuel consumption and 11% in injection pressure are observed at idle operation between B0 and B10. A reduction of the CO and HC emissions is also registered with a maximum variation at the cruise regime (80% Revolutions Per Minute—RPM).

Keywords: biodiesel; gas turbine; turbojet; energy performance; emissions; aviation

1. Introduction

Aviation transport is an important energy consumer sector that strongly depends on fossil fuels such as jet/kerosene, diesel, and gasoline. It is also largely vulnerable to the availability and cost of fossil fuel [1]. Domestic and international aviation transport consumed almost 3.2% of the world's energy consumption in 2015 [2]. Moreover, fossil fuel prices strongly affect the sustainability of the aviation sector, projecting a rise of jet fuel costs near $100 USD/ton per year [3]. In 2016, air transportation contributed around 2.3% of the global anthropogenic CO_2 emissions [4], significantly contributing to the greenhouse as (GHG) emissions. As a result of the increasing freight transport activities throughout the world and low utilization of renewable energies in the aircraft powertrain splits, the international-aviation sector is constantly evaluating a variety of alternatives for the replacement of conventional fuels with alternative fuels [3].

Alternative aviation fuels can be produced from a variety of renewable biomass feedstock as soybean, palm, coconut, or jatropha with different production routes as biomass to liquid (BTL), Fischer Tropsch, hydro-processing, biochemical fermentation, or transesterification. The mixtures of fatty acid methyl esters (FAME) derived from vegetable oils or biodiesel have shown advantages in miscibility with fossil fuels and contain no sulfur or aromatic compounds [3]. Biodiesel has been studied worldwide as an alternative fuel for turbine engines, in part, explained by the current commercial availability of this biofuel. The substitution of biodiesel has been studied not only for generating thrust in aircrafts [5–7] but also for electric power generation systems [8–10].

According to the literature, one main difference between biofuel and fossil fuel is the higher viscosity of the former. For example, when using a mixture of soybean biodiesel with ultra-low-sulfur-diesel, the viscosity difference affects the atomization of fuel in the nozzle [11]. Several studies of biodiesel blends used in aviation engines have shown no significative detrimental effect on the engine performance compared to the use of pure jet fuels [7,12]. The exhaust gas emissions expose no significant variations of O_2 and CO_2 compounds [13], but a significant rise in CO emissions is reported when increasing the biodiesel content as a result of an inadequate atomizing of the fuel that causes a more incomplete combustion [6,14]. There are still several challenges to overcome with the utilization of biodiesel in aviation turbines regarding the high cost of fuel production, the degradative effect of biofuels over fuel systems linked to the crucial operating safety, and the adjustment of the biodiesel properties to aviation standards [15].

Unfortunately, the experimentation on full scale turbines—which may clarify the acceptable operation parameters—are expensive and complex, limiting most of the reported experimentation to small size turbojets ranging from 10 to 50 kW [16]. Studies reported by Chiaramonti et al. [7] and Nascimento et al. [6] evaluated the effect of blends of diesel and biodiesel on the energy performance and the emissions of a micro-turbine with energy power ratings of 20 and 30 kW, indicating no changes in the energy performance of the engine. In opposition, French [17] and Habib et al. [18] report a reduction in the thrust force by around 8% and, thus, a reduction in the thrust specific fuel consumption when mixing biodiesel and Jet A in 30 kW micro-turbines. Regarding the emission, a reduction in the evolved CO, HC, and NOx is noted by Chiaramonti et al. [7], but a formation of cloudiness is reported by French [17].

On the other hand, the evaluation of turbines with an energy power rating above 100 kW is reported in a few studies, except in the ones presented by Corporan et al. [5] and Lupandin et al. [19]. In this sense, Corporan et al. [5] tested mixtures of JP-8 and up to 20% volume content (*v*/*v*) of biodiesel in a 230 kW helicopter engine ref. Allison T63-A-700, reporting an increase of 20% in the particulate emissions at the idle regime. In that study, a rise of 4% in fuel consumption was reported due to inaccuracy in the fuel control unit (FCU). Lupandin et al. [19] tested a 2.5 MW power generation gas turbine ref. GT2500 using mixtures of Jet A and up to 12% *v*/*v* of biodiesel. The experimentation concluded an increase in the exhaust gas temperature, fuel consumption, and the evolved CO emissions, also exposing the plugging of biodiesel in the filtration system.

The former incongruences in the behavior of each micro-turbine preclude an accurate prediction of the effect of biodiesel in large-scale turbines. Furthermore, the lack of experimental information and validation in real aeronautic turbines with biodiesel is remarkable, with the major absence of information on the effects in the operating performance, emissions, and in the mechanical wearing of components. The main purpose of the current study is to present the experimental methodology and facility in a 640 kW full-scale J69-T-25A turbojet engine operated at different load regimes using mixtures of Jet A1 with 0%, 5%, and 10% oil palm biodiesel. We describe the test facility and experimental procedures, including the test bench and monitoring system, and the experimental planning. The main findings are related to the overall engine operating performance and the exhaust emissions by measuring the: thrust force (T), fuel mass flow ($\dot{m}_f$), and volume fractions of O_2, CO_2, CO, and HC. The present work is complementary to computational analysis of the performance of gas turbine engines operating with biodiesel blends, such as the one presented in [20].

2. Materials and Methods

Two fuels were used in the present study: a commercial kerosene-based Jet A1, and a Colombian palm oil biodiesel with a commercial reference BioD, Premium 360. The experimental plan was conducted with fuel blends of Jet A1 and biodiesel using the biodiesel volume contents (w_{BD}) of 0% (B0), 5% (B5), and 10% (B10). The blends were prepared in a tank with a nominal capacity of 4 m^3, using a transfer pump and a flow meter Tuthill Fill-Rite model 820 (accuracy of 0.1%). An initial volume $V_{i,B0} = 1762 \pm 8$ L of Jet A1 was filled in the tank and was consumed in the baseline experiments B0 until a final volume $V_{f,B0}$ was obtained. Thereafter, a volume of biodiesel was added to the tank to increase the biodiesel volume fraction. The same procedure was repeated for the mixture of B10. The experimental level and the accuracy of the volume contents of biodiesel were always kept below 2.5%. These fuel blends were used to test the engine performance operating at different load regimes from idle to maximum throttle.

2.1. Experimental Facility and Procedure

The experimental facility employed in this study is presented in Figures 1 and 2. The tested engine is a turbojet engine Teledyne Continental Aviation and Engineering—CAE J69 used in the military aircraft Cessna T-37 by the Colombian Air Force. The engine was installed on a testing bench, as presented in Figure 1a, and the instrumentation used corresponds to the one described in the engine technical manual (T.O. 2J-J69-72) [21]. The variables: Outside Air Temperature OAT, Exhaust Gas Temperature—EGT, rotational speed of the main shaft—ω_{shaft}, thrust force—T, and fuel injection pressure—P_f were measured and acquired in the control booth every 5 s only using a National Instruments ref. cRIO data logger.

The main components of the experimental facility are schematically presented in Figure 2a, consisting of the turbojet engine, the control and instrumentation sets, the thrust force measurement system, the fuel delivery system, the exhaust gas sampling stand, and the lubrication system. The Teledyne CAE J69 is a turbojet engine of model variant J69-T-25A with its main technical specifications, as indicated in Table 1 and with a nominal energy power rating of 640 kW. The rotational speed of the main shaft (ω_{shaft}) is measured with an AAE model MS25038-4 tachometer generator, with a speed ratio, measurement range, and accuracy of 6:1, 0–4200 RPM, and 1%, respectively.

(a)

(b)

Figure 1. (**a**) General aspect of the J69 engine installed in the testing bench; (**b**) detail of the control booth used in the experimentation.

The exhaust gas temperature (EGT) is one main operative parameter that allows the evaluation of the mechanical and energetic behavior of the engine, as this is affected by the combustion conditions in the hot section of the turbine, the changes of the air to fuel ratio (AFR) at the different operating regimes, and the final power conversion and heat dissipation of the turbine. The outside air temperature (OAT) affects the density of the intake air, modifying the compression efficiency and the general performance of the engine. The values of OAT and EGT were measured with a set of 6 thermocouples, ref. 700530, normally used during a flight routine, which are installed at the air inlet duct and the outer tailpipe, as presented in Figure 2a. The layout of the installation of these thermocouples is presented in Figure 2c in accordance with the standard instrumentation of the engine reported in its technical manual [21].

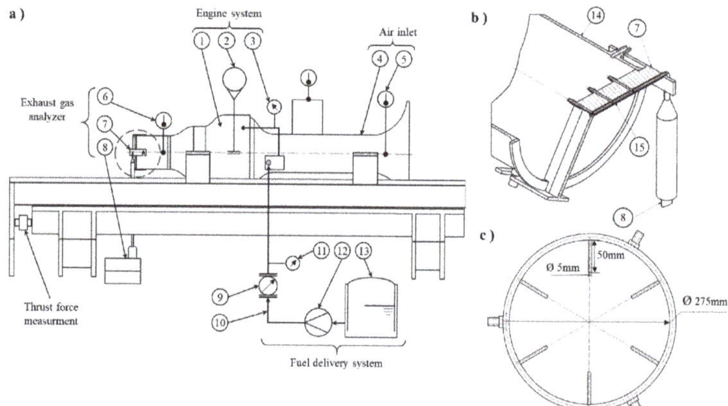

Figure 2. (**a**) General J69 engine experimental facility detailing peripheral components and instrumentation: (1) turbojet engine model J69T-25A, (2) tachometer, (3) fuel pressure manometer, (4) air inlet duct, (5) air inlet temperature probe set, (6) exhaust gas temperature (EGT) thermocouple set, (7) exhaust gas sampling rake, (8) exhaust gas analyzer, (9) fuel flow meter, (10) fuel supply line, (11) fuel inlet pressure manometer, (12) fuel supply pump, and (13) fuel tank; (**b**) detail of the exhaust gas sampling rake: (14) outer tailpipe, (15) gas sampling ducts; (**c**) the layout of the installation of the thermocouples, in accordance with the standard instrumentation of the engine technical manual [21]. (Source: Author's own figures).

Table 1. General specifications and operating environment of the J69 engine.

General Specifications of the J69 Engine	
Engine reference	Teledyne CAE J69
Variant	J69-T-25A
Engine type	Turbojet—Single spool
Rotational speed max./RPM	21,730
Thrust max./N	4560
EGT max./°C	663
Compressor type	Centrifugal—1 stage
Inlet air flow max./kg/s	9.07
Pressure ratio max.	3.9
Turbine type	Axial flow—1 stage
Fuel distributor	Centrifugal/Slinger holes
No. Of fuel injectors	2
No. Of Slinger Holes	16
Fuel type	JP-4, Jet A, Jet A1
Fuel consumption max./g/s	143.6
Operating Environment	
Location	Madrid, Colombia
MSN/m	2554
OAT/°C	16–26
RH/%	25–56
P_{atm}/hPa	1017

The exhaust gas is sampled at the outer tailpipe with an exhaust gas sampling rake, as presented in Figure 2b, designed to meet the standards of the International Civil Aviation Organization [22]. The volume fractions of O_2, CO_2, CO, and HC at the exhaust are measured using a Hanatech ref. IM2400 Ultra 4/5 gas analyzer with ranges 0%–23%, 0%–20%, 0%–10%, and 0–10,000 ppm for the O_2, CO_2, CO, and HC, respectively.

The fuel mass flow ($\dot{m}_f$) was measured using a Tuthill Fill-Rite model 820 digital flow meter, ranging from 7.6 L/min to 75.7 L/min with an accuracy of 0.5%. To sense the fuel injection pressure (P_f), a Sunpass 300 PSI pressure gage model with an accuracy of 0.4% was used. During the experimentation, the fuel blends were stored in the tank trailer and pumped with an external transfer pump (with a flow range from 20 L/min to 120 L/min and 345 kPa of max. pressure) into the fuel pump group (FPG). In the FPG, the fuel was pressurized up to 1448 kPa and pumped into the fuel control unit (FCU), where the fuel mass flow was controlled as a function of the acceleration control, ω_{shaft}, EGT, and the sensed pressure of the pre-heated primary air.

The thrust force (T) was measured with a set of two Toroid Corp. 3965 model load cells (with a load range of 11,340 kg and an accuracy of 1%). General specifications of the experimental facility and the operating conditions of the engine are presented in Table 1. Experimentation was carried out in the atmospheric environment of Madrid, Colombia. The barometric pressure, the temperature, and relative humidity of the ambient air were measured with a Wallace and Tiernan ref. FA 112170 barometer and an Omet ref. C3121 thermo-hygrometer.

To run an experiment, the testing bench sensors are first checked, and the fuel line coming from the tank trailer is pressurized. Then, the turbine shaft is actioned to rotate to 8257 RPM using a starter-generator drive. The fuel blend is injected with 2 sets of starting fuel nozzles. Thereupon, a set of 2 spark plugs are used to ignite the combustion, and the fuel distributor starts the fuel injection. At this procedure, the exhaust gas temperature immediately reaches the max. operation value. The ignition procedure checks if the value of EGT is below 662 °C, such that the starter-generator drive and the spark plugs are switched off, and the engine is left over at the idle regime (8257 RPM). Once the engine is turned on and all the peripheral systems are checked, the engine is gradually sped up to the take-off regime (21,730 RPM) until the lubrication oil temperature reaches from 37 °C to 65 °C. The response variables are registered for 5 min at steady-state conditions.

Then, the engine is slowed down to the cruise regime (17,384 RPM), left over for 5 min to attain steady-state conditions, and the response variables are again registered for another 5 min. The register procedure is repeated for the 15,211 RPM and the idle (8257 RPM) regime. Lastly, the engine is left over at the cruise regime for 10 min to evaluate changes in operation variables. To finish the experimental run, the engine is slowed down gradually to the idle regime, and the combustion systems are turned off. The engine is then cooled down until EGT is below 120 °C, and the main components of the engine are visually checked. After that, a new fuel blend is prepared and tested by repeating the same procedure. Three test replicas were done for each blend to have a statistical control of the experimental measurements.

2.2. Experimental Planning and Data Processing

The experiments are codified with "B" for w_{BD} and "RPM" for the engine operating regime as a percentage ratio of the maximum ω_{shaft} of 21,730 RPM. The baseline experiments with no biodiesel content (B0-RPM38, B0-RPM70, B0-RPM80, B0-RPM100) were selected to evaluate the design operating regimes of the engine using pure Jet A1. The biodiesel volume contents of B5 and B10 were selected to evaluate the effect of the biodiesel substitution in the engine operation as no prior evidence of operation with biodiesel is reported in the literature for the J-69 engine. The operating regimes of the engine were set to 38% RPM (8257 RPM), 80% RPM (17,384 RPM), and 100% RPM (21,730 RPM) in order to evaluate the idle, cruise, and take-off regimes, respectively. An additional operating regime of 70% RPM (15,211 RPM) was selected to assess the effect of the biodiesel substitution in an intermediate regime between the idle and cruise regimes.

These regimes were chosen to be evaluated since they are the most frequently used during a flight. The idle regime (38% RPM) is the stabilization condition after the start-up of the engine with no acceleration. At this regime, all of the operating parameters of the engine are normally verified before the flight to avoid overheating, possible fuel and oil leakages, or excessive vibration. The cruise regime (80% RPM) and intermediate regime (70% RPM) are commonly used during flight but also to evaluate the performance of the engine during the taxing of the aircraft at minimum thrust. Finally, the take-off regime (100% RPM) is used during the take-off and landing of the aircraft, normally reporting the maximum power generation of the engine with maximum fuel consumption and mechanical stresses of the main components.

During the steady-state of the engine at every operating regime, the OAT, EGT, ω_{shaft}, T, and P_f variables were measured and automatically acquired every 5 s by a National Instruments ref. cRIO data logger. On the other hand, the value of $\dot{m}_f$ and the volume fractions of O_2, CO_2, CO, and HC at the exhaust gas were manually registered every 15 s by using the perimetric instrumentation (defined in Figure 2a) only during the steady-state regimes. To test repeatability, the experimental runs were executed three times.

3. Results and Discussion

Figures 3 and 4 present the results of the measured variables T, EGT, $\dot{m}_f$, and P_f following the experimental plan, compiling, at first, the baseline experiments (B0) with non-biodiesel content and the substitution of biodiesel B5 and B10, varying the engine responses at the operating regimes: idle—38% RPM, 70% RPM70; cruise—80% RPM; take-off—100% RPM. The obtained max. coefficients of variations are as follows: T is 4.2%, EGT is 0.4%, $\dot{m}_f$ is 2.3%, and P_f is 2.5%. Hence, the extracted results and discussions rely on statistically representative data. The minimum and maximum values of T, $\dot{m}_f$, and P_f are attained at idle and cruise regimes, respectively. The minimum EGT value occurs between 70% RPM and cruise regimes, as a result of the optimal heat dissipation of the engine [21].

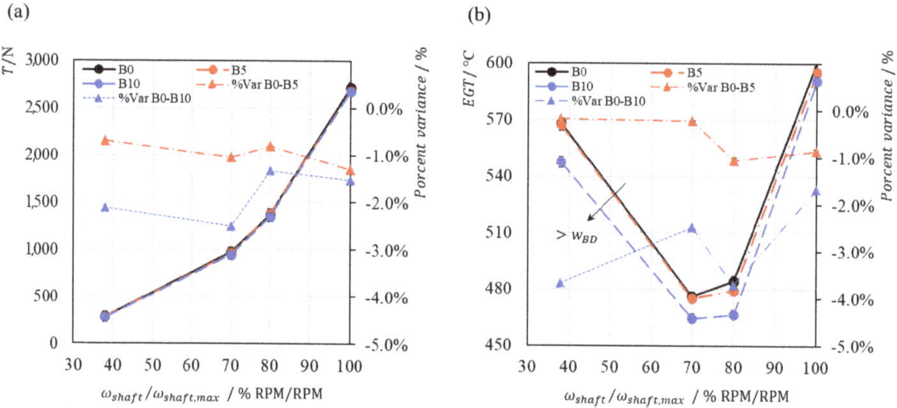

Figure 3. Effect of the biodiesel content for experiments B0, B5, and B10 varying the engine regime ($\omega_{shaft}/\omega_{shaft,max}$) in the: (**a**) thrust force (T) and the absolute percent variance, (**b**) exhaust gas temperature (EGT) and the absolute percent variance.

The effect of the biodiesel content increases up to 10% *v/v* and does not reveal a significant variation in the T and EGT variables for every operation regime as the absolute percent variation is below 2.5% and 3.8%, respectively. On the other hand, a significative statistical difference is found at the idle regime for $\dot{m}_f$ and P_f, with a level of significance of 90%. Instead, no significative differences are found for the 70% RPM, cruise, and take-off regimes. The rise in biodiesel content from 0% to 10% increases the $\dot{m}_f$ by 36% and P_f by 11% at the idle operation, mainly explained by a fuel control unit

(FCU) malfunctioning as a result of lower energy content of the biodiesel, poor atomization of the fuel, and possible saturation of the fuel filters.

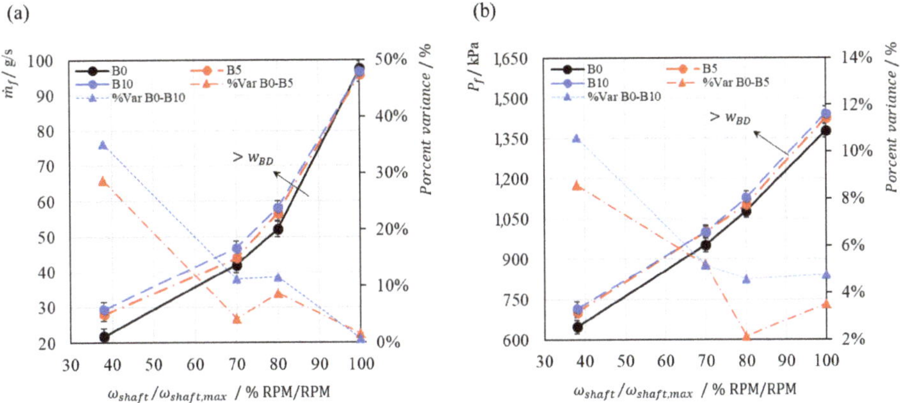

Figure 4. Effect of the biodiesel content for experiments B0, B5, and B10 varying the engine regime ($\omega_{shaft}/\omega_{shaft,max}$) in the: (**a**) fuel consumption ($\dot{m}_f$) and the absolute percent variance, (**b**) fuel injection pressure (P_{fuel}) and the absolute percent variance.

The FCU is a hydro-mechanical component that dozes the amount of fuel by modifying the valves opening in a set of needles as a function of the ω_{shaft}, T_{air}, and the density and viscosity of the fuel blend. Thus, modification in the hydro-mechanical properties of the fuel blend alters the FCU operation, increasing the stress of the needle valves and the fuel injection pressure. Hoxie et al. [11] and Corporan et al. [5] also reported those variations in $\dot{m}_f$ and P_f, attributed to the decrease of the fuel blend HHV_{ad} that reduces the rate of energy supplied by the fuel and to poor atomization of the fuel in the combustor chamber when increasing the biodiesel content. According to Lupandin et al. [19], an increase in the viscosity of the fuel mixture and the plugging of biodiesel in the filtration system was registered during those previous experimentations.

Also, the emission measurements of O_2, CO_2, CO, and HC are presented in Figures 5 and 6. The max. standard deviation for the volume fraction of exhaust gases is as follows: O_2 is 0.13%, CO_2 is 0.04%, CO is 0.04%, and HC is 2.5 ppm. A maximum volume fraction in the flue gas of O_2 and a minimum of CO_2 are observed at the cruise regime (80% RPM), indicating an increase of the air to fuel ratio (AFR). Prior variations are also reported by Chiong et al. [13] and Rochelle et al. [6,14], in which a higher oxygen content was linked to biodiesel compared to fossil fuel, increasing the percent excess combustion air and thus increasing the AFR. Nonetheless, the evolution of CO and HC presented in Figure 6 exposes a reduction in the volume fraction when increasing the main shaft rotational speed with a minimum value of CO at the take-off regime (100% RPM). Moreover, no HC emissions are observed at take-off as a result of a possible increase in the excess air of combustion.

A reduction of the CO and HC emissions is also presented in Figure 6 with the rise of biodiesel content up to 10% *v*/*v*, with a maximum variation between 70% RPM and the cruise regime (80% RPM). The reduction of the HC emissions when using 10% *v*/*v* near 58% indicates a higher combustion efficiency and a more complete combustion within the engine. Kimble et al. [23] and Chiaramoniti et al. [7] also reported a reduction in the evolved CO when increasing the atomizing pressure from 0.3 bar to 1.1 bar, leading to a better flame stability within the combustion chamber and raising the fuel injection temperature above 120 °C, as the fuel viscosity is reduced when increasing the injection temperature and thus, improves the atomization and volatilization processes.

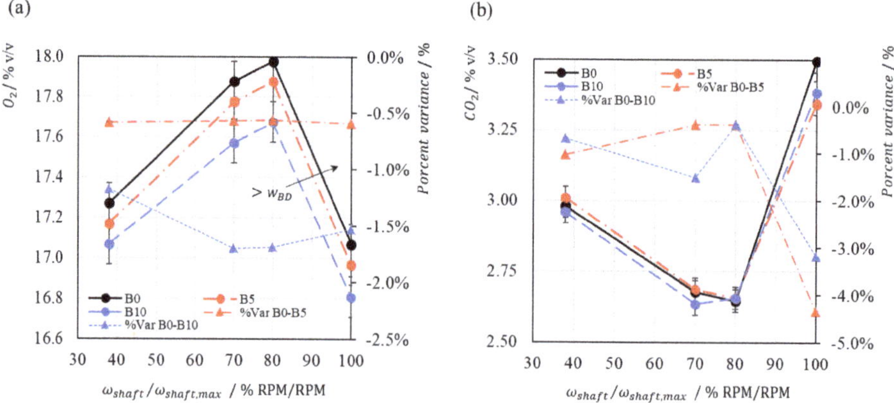

Figure 5. Effect of the biodiesel content for experiments B0, B5, and B10 varying the engine regime ($\omega_{shaft}/\omega_{shaft,max}$) in the: (**a**) content of oxygen in flue gas (O_2) and the absolute percent variance, (**b**) content of carbon dioxide in flue gas (CO_2) and the absolute percent variance.

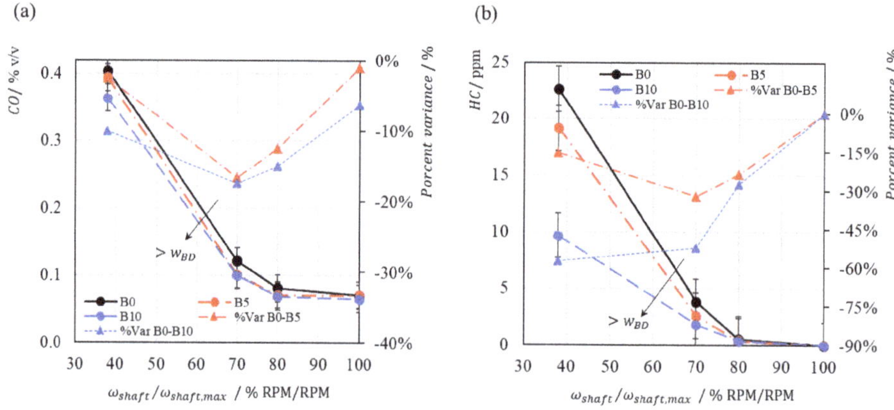

Figure 6. Effect of the biodiesel content for experiments B0, B5, and B10 varying the engine regime ($\omega_{shaft}/\omega_{shaft,max}$) in the: (**a**) content of carbon monoxide in flue gas (CO) and the absolute percent variance, (**b**) hydrocarbons in flue gas (HC) and the absolute percent variance.

The effects of these changes in the emission can be further identified as a possible variation of the AFR and the *EA* at different biodiesel contents and operating regimes. The *EA* tends to decrease when increasing the biodiesel content from 0% to 10%, as the AFR_{st} is reduced. Moreover, the additional excess air and the slight reduction of the EGT are indicators of a possible higher influence of the secondary air stream during combustion. Even though prior behavior can be linked to the variation of the AFR and the excess air of combustion, no measurement of the airflow intake of the engine at all the operating regimes precludes the validation of the effect of those variables in the measured emissions.

4. Conclusions

The testing of the turbine J69-T-25A using blends of Jet A1 and biodiesel validates the feasibility of using up to 10% *v/v* of oil palm biodiesel in a full-size military aviation engine. The experimental methodology and the used facility allowed the measurement of operating parameters of the engine. No significant variations in the thrust force and EGT were observed in all the operation regimes with the utilization up to 10% *v/v* of biodiesel. Nonetheless, a 36% increase in fuel consumption and an 11% increase in injection pressure were observed at idle operation between B0 and B10.

Additionally, the perimetral instrumentation allowed the quantification of the evolved emissions in terms of the volume content of CO and HC in the flue gas. At the cruise regime (80% RPM), a maximum reduction of the CO and HC emissions was observed with a maximum variation of 25% and 58%, respectively. The reduction of those emissions indicates a sustainable operation of the J69 engine with 10% of biodiesel content by presenting a more complete combustion with a significative reduction in the operating performance variables.

In further studies, a more detailed examination of the energy performance in terms of energy efficiency, thrust-specific fuel consumption, and combustion reaction is recommended. Furthermore, the validation of a higher substitution of biodiesel in the fuel blend is advised in order to identify possible detrimental aspects of the operating performance or mechanical wearing of the main components.

Author Contributions: Conceptualization, G.T., V.S., J.P. and M.L.; data curation, G.T., G.M. and M.G.; formal analysis, G.T. and C.B.-R.; funding acquisition, M.L.; investigation, G.T., G.M. and V.S.; methodology, G.T., G.M. and V.S.; project administration, V.S. and M.L.; resources, G.M., M.G., J.P. and M.L.; supervision, C.B.-R., J.P. and M.L.; visualization, G.T.; writing—original draft, G.T., G.M. and V.S.; writing—review and editing, C.B.-R.

Funding: This work was supported by COLCIENCIAS (Grant No. 56752). The Colombian Air Force—FAC and The Universidad ECCI.

Acknowledgments: The authors would like to thank Hamilton Henao from CAMAN for all the support during the experimental labor; Nelson Jimenez from ESUFA; and M.G., Renso Arango from the Universidad ECCI.

Conflicts of Interest: The authors declare no conflict of interest.

Abbreviations

The following nomenclature is used in this manuscript:

Symbol	Definition	Units
$\%EA$	Percent excess combustion air	% g/g
P_f	Fuel pressure	kPa
$\dot{m}_f$	Fuel mass flow	g/s
w_{BD}	Biodiesel volume content in fuel blend	% *v/v*
ω_{shaft}	Main shaft rotational speed	RPM
AFR	Air/fuel ratio	g/g
EGT	Exhaust gas temperature	-
HHV	Higher heating value	MJ/kg
T	Thrust force	N
OAT	Outside Air Temperature	°C

References

1. Abu Talib, A.R.; Gires, E.; Ahmad, M.T. Performance Evaluation of a Small-Scale Turbojet Engine Running on Palm Oil Biodiesel Blends. *J. Fuels* **2014**. [CrossRef]
2. Teske, S. *Achieving the Paris Climate Agreement Goals: Global and Regional 100% Renewable Energy Scenarios with Non-Energy GHG Pathways for +1.5 C and +2 C*; Institute for Sustainable Futures University of Technology Sydney, Ed.; Springer: Sydney, NSW, Australia, 2019; ISBN 9783030058432.
3. Kandaramath Hari, T.; Yaakob, Z.; Binitha, N.N. Aviation biofuel from renewable resources: Routes, opportunities and challenges. *Renew. Sustain. Energy Rev.* **2015**, *42*, 1234–1244. [CrossRef]
4. Coban, K.; Colpan, C.O.; Karakoc, T.H. Application of thermodynamic laws on a military helicopter engine. *Energy* **2017**, *140*, 1427–1436. [CrossRef]
5. Corporan, E.; Reich, R.; Larson, V.; Aulich, T.; Mann, M.; Seames, W. Impacts of biodiesel on pollutant emissions of a JP-8–fueled turbine engine. *J. Air Waste Manag. Assoc.* **2005**, *55*, 940–949. [CrossRef] [PubMed]
6. Nascimento, M.A.R.; Lora, E.S.; Corrêa, P.S.P.; Andrade, R.V.; Rendon, M.A.; Venturini, O.J.; Ramirez, G.A.S. Biodiesel fuel in diesel micro-turbine engines: Modelling and experimental evaluation. *Energy* **2008**, *33*, 233–240. [CrossRef]

7. Chiaramonti, D.; Rizzo, A.M.; Spadi, A.; Prussi, M.; Riccio, G.; Martelli, F. Exhaust emissions from liquid fuel micro gas turbine fed with diesel oil, biodiesel and vegetable oil. *Appl. Energy* **2013**, *101*, 349–356. [CrossRef]
8. Sundararaj, R.H.; Dinesh, R.; Kumar, A.; Sekar, T.C.; Pandey, V.; Kushari, A.; Puri, S.K. Combustion and emission characteristics from biojet fuel blends in a gas turbine combustor. *Energy* **2019**, *182*, 689–705. [CrossRef]
9. Udeh, G.T.; Udeh, P.O. Comparative thermo-economic analysis of multi-fuel fired gas turbine power plant. *Renew. Energy* **2019**, *133*, 295–306. [CrossRef]
10. Harahap, F.; Silveira, S.; Khatiwada, D. Cost competitiveness of palm oil biodiesel production in Indonesia. *Energy* **2019**, *170*, 62–72. [CrossRef]
11. Hoxie, A.; Anderson, M. Evaluating high volume blends of vegetable oil in micro-gas turbine engines. *Renew. Energy* **2017**, *101*, 886–893. [CrossRef]
12. Allouis, C.; Beretta, F.; Minutolo, P.; Pagliara, R.; Sirignano, M.; Sgro, L.A.; Anna, A.D. Measurements of ultrafine particles from a gas-turbine burning biofuels. *Exp. Therm. Fluid Sci.* **2010**, *34*, 258–261. [CrossRef]
13. Chiong, M.C.; Chong, C.T.; Ng, J.H.; Lam, S.S.; Tran, M.V.; Chong, W.W.F.; Mohd Jaafar, M.N.; Valera-Medina, A. Liquid biofuels production and emissions performance in gas turbines: A review. *Energy Convers. Manag.* **2018**, *173*, 640–658. [CrossRef]
14. Rochelle, D.; Najafi, H. A review of the effect of biodiesel on gas turbine emissions and performance. *Renew. Sustain. Energy Rev.* **2019**, *105*, 129–137. [CrossRef]
15. Čerňan, J.; Hocko, M.; Cúttová, M. Safety risks of biofuel utilization in aircraft operations. *Transp. Res. Procedia* **2017**, *28*, 141–148. [CrossRef]
16. Badami, M.; Nuccio, P.; Signoretto, A. Experimental and numerical analysis of a small-scale turbojet engine. *Energy Convers. Manag.* **2013**, *76*, 225–233. [CrossRef]
17. French, K.W. Recycled fuel performance in the SR-30 gas turbine. In Proceedings of the 2003 American Society for Engineering Education Annual Conference Exposition, Nashville, TN, USA, 22–25 June 2003.
18. Habib, Z.; Parthasarathy, R.; Gollahalli, S. Performance and emission characteristics of biofuel in a small-scale gas turbine engine. *Appl. Energy* **2010**, *87*, 1701–1709. [CrossRef]
19. Lupandin, V.; Thamburaj, R.; Nikolayev, A. Test results of the OGT2500 Gas Turbine. Engine Running on Alternative Fuels: BioOil, Ethanol, BioDiesel and Crude Oil 2005. In Proceedings of the ASME Turbo Expo 2005. Power for Land, Sea, and Air, Reno, NV, USA, 6–9 June 2005; pp. 421–426.
20. Bayona-Roa, C.; Solís-Chaves, J.; Bonilla, J.; Rodriguez-Melendez, A.; Castellanos, D. Computational Simulation of PT6A Gas Turbine Engine Operating with Different Blends of Biodiesel—A Transient-Response Analysis. *Energies* **2019**, *12*, 4258. [CrossRef]
21. Teledyne Continental Aviation and Engineering—CAE. *Technichal Manual T.0. 2J-J69-72 Intermediate Maintenance Instructions Turbojet Engine Model No. J69-T-25-A*; US Air Force: Oklahoma City, OK, USA, 2004.
22. Choi, S.; Lee, D.; Park, J. Ignition and combustion characteristics of the gas turbine slinger combustor. *J. Mech. Sci. Technol.* **2008**, *22*, 538–544. [CrossRef]
23. Kimble-Thom, M.A.; Stanley, D.L.; Cholis, J.T.; Lopp, D.W. The Use of Bio-Fuels as Additives and Extenders for Aviation Turbine Fuels. In Proceedings of the ASME 1999 International Gas Turbine and Aeroengine Congress and Exhibition, Indianapolis, IN, USA, 7–10 June 1999; ASME: New York, NY, USA, 2014. V002T01A007.

Article

Experimental Analysis of an Air Heat Pump for Heating Service Using a "Hardware-In-The-Loop" System

Paolo Conti, Carlo Bartoli, Alessandro Franco and Daniele Testi *

Department of Energy, Systems, Territory, and Constructions Engineering (DESTEC), University of Pisa, Largo Lucio Lazzarino, 56122 Pisa, Italy; paolo.conti@unipi.it (P.C.); carlo.bartoli@unipi.it (C.B.); alessandro.franco@unipi.it (A.F.)

* Correspondence: daniele.testi@unipi.it

Received: 10 July 2020; Accepted: 18 August 2020; Published: 1 September 2020

Abstract: Estimating and optimizing the dynamic performance of a heat pump system coupled to a building is a paramount yet complex task, especially under intermittent conditions. This paper presents the "hardware-in-the-loop" experimental campaign of an air-source heat pump serving a typical dwelling in Pisa (Italy). The experimental apparatus uses real pieces of equipment, together with a thermal load emulator controlled by a full energy dynamic simulation of the considered building. Real weather data are continuously collected and used to run the simulation. The experimental campaign was performed from November 2019 to February 2020, measuring the system performances under real climate and load dynamics. With a water set point equal to 40 °C, the average heat pump coefficient of performance was about 3, while the overall building-plant performance was around 2. The deviation between the two performance indexes can be ascribed to the continuous on-off signals given by the zone thermostat due to the oversized capacity of the heat emission system. The overall performance raised to 2.5 thanks to a smoother operation obtained with reduced supply temperature (35 °C) and fan coil speed. The paper demonstrates the relevance of a dynamic analysis of the building-HVAC system and the potential of the "hardware-in-the-loop" approach in assessing actual part-load heat pump performances with respect to the standard stationary methodology.

Keywords: hardware-in-the-loop; heat pumps; dynamic simulation; experimental performances; control strategy; partial loads; on-off cycles; building dynamics; building-heating system coupling

1. Introduction

Heat Pump (HP) is a key technology to increase the energy efficiency and the use of renewable energy in civil and residential buildings [1]. With respect to electric or natural-gas boilers, HPs have higher energy conversion efficiency and allow the exploitation of the renewable energy coming from the outdoor source (e.g., external air), together with the renewable share of the electricity production mix. In this regard, heat pumps are recognized to be increasingly important for electrical grids (micro- and smart-included), as well as for distributed energy systems that aim at even higher renewable energy contribution [1–6]. The fluctuating and stochastic production of renewables represents a critical issue to ensure the balance between power production and load: HPs may mitigate this drawback by shifting a given share of the thermal energy demand to the electrical one [5,6]. A proper cooperation among HPs, back-up generators and energy storages may additionally increase the flexibility of the overall energy system [4,7]. Demand response techniques are also possible [8,9].

Several scientific and technological issues in the field of HP systems are still unsolved. One of those is the problem of the efficiency reduction under dynamic operating conditions [10]. Especially in mild climates conditions and during intermediate seasons, the actual thermal load profile results are notably lower than the design capacity and intermittent operation occurs. The frequent on–off cycles and the resulting dynamic regime for the whole HP system determine a considerable reduction in the mean (or seasonal) Coefficient of Performance (*SCOP*): while in nominal operating conditions the *SCOP* ranges from values of 3 (in cooling mode) to 4 (in heating mode), in intermittent operation the values are often reduced even below the values of 2 and 3 [1,11].

The development of sizing and management strategies is fundamental to improve the actual performances of HP-based systems and guarantee low energy consumption, reduced operating costs, and high comfort levels for occupants. From this perspective, some works based on the dynamic simulation of the Heating and Ventilation Air Conditioning (HVAC)-building system have been proposed in the literature [6,12]. However, the results obtained through a simulation approach would always be followed by some experimental tests in order to validate assumptions, subsystems models, dynamics timescales, and estimated operative performances [13–15]; otherwise, results can be affected by notable errors or correspond to impracticable control actions to be realized in a real system [1].

Full experimental campaigns are often unfeasible for building energy systems, due to the long time required, technical and economical efforts. Then, the use of "hardware-in-the-loop" (HiL) systems is a convenient experimental approach that is quite novel for these applications. In other fields, such as automotive and aerospace industries, HiL tests are standard practice and currently their importance is even increasing, due to the onset of advanced driver assistance and automated driving [16]. In the case of building energy systems, with the HiL approach, the experimental campaign is performed in a laboratory environment through an apparatus made of real thermal generators, heat storages and/or any other pieces of equipment, while the real building is replaced by a thermal load emulator controlled by a full energy dynamic simulation [15,17–19]. Other technical units may also be substituted by appropriate emulators, to be integrated with the energy system. Using real weather data, the tested devices (e.g., the heat pump) experience the same working conditions and load profiles that they would experience in a real application [9,10,13]. HiL experimentation is also characterized by a great flexibility, because it can test and optimize many different building-HVAC configurations and control strategies [17,20,21].

In this experimental framework, the objective of the present paper is to show the relevance of the dynamic coupling between building and heat pump system in terms of energy performance, especially when frequent on–off cycles occur, and quantitatively assess the deviations from standard methods based on stationary measurements.

The conceptual layout of the building emulator and thermo-hydraulic HiL experimental system available in the laboratories of DESTEC of the University of Pisa, named Building Emulator and Integrated Energy System (BE-IES), is shown in Figure 1, including all the installed devices, measurements, controls, and corresponding energy fluxes.

Following a thorough description of the BE-IES laboratory components and management options, the results of an experimental campaign concerning the dynamic operation of an air-source HP serving a typical apartment located in Pisa in wintertime are presented and analyzed. In particular, the paper focuses on the effects of the supply temperature to the heat emission system and the resulting on–off signal provided by the typical zone thermostats located inside the building.

Experimental performances are compared with the estimated ones according to current widespread methodologies based on the interpolation of manufacturers' experimental data [22–24]. These methods can be classified as "quasi-stationary", as they are based on a sequence of steady-state working conditions; the effects of dynamic phenomena (e.g., on–off cycles) are evaluated through empirical penalization factors. However, these inefficiencies can be relevant and hard to model with simplified correlations as they depend on specific load profile, heating loop characteristics, HVAC-building interconnected dynamics; therefore, the application of standardized methodologies can lead to wrong estimations of the system efficiency and related economic viability.

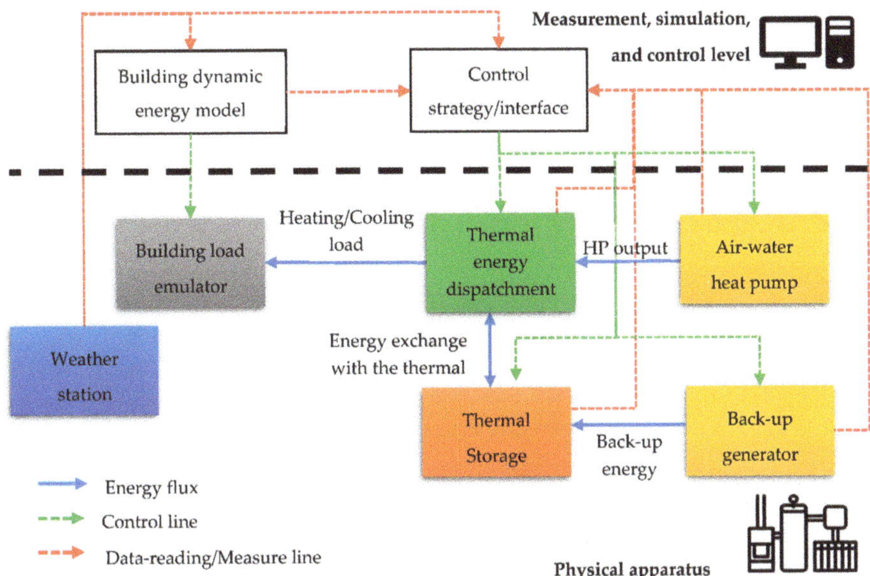

Figure 1. Components and logical scheme of the building emulator and integrated energy system (BE-IES) "hardware-in-the-loop" apparatus.

The rest of the paper is structured as follows: Section 2 describes the BE-IES experimental apparatus and the operational strategy of the HiL system; Section 3 presents the dynamic energy model of the building and heat emission system that control the building emulator; Section 4 presents the case study and the three tested operational strategies. Finally, Section 5 discusses the experimental data, the actual coefficient of performances of the building-HVAC system with respect to the corresponding ones evaluated through state-of-the-art technical standards (briefly recapped in Appendix A), and analyzes the causes of the thermal inefficiencies connected with the operation of the heat pump and the overall HVAC system, leading to the conclusions of Section 6.

2. Description of the "Hardware-In-The-Loop" Apparatus

The experimental apparatus is made of several thermal and electrical devices: some of them are real HVAC piece of equipment, others act as emulators (e.g., the building). The apparatus sizing was tailored to the typical design thermal load of single-family houses, namely some kW.

2.1. Equipment Description and Layout

The list of components and the overall layout of the experimental apparatus are shown in Table 1 and Figure 2, respectively. The system has separate generation and user sections, connected by the main heat exchanger (mHEx). The building emulator (BE), the mHEx and the thermal storage (TS) are connected through two three-way control valves that link two devices at a time, according to the signal received by the main monitoring and control (M&C) system. The apparatus can thus operate in three possible configurations:

(a) AHP-TS: the heat pump heats up the thermal storage;
(b) mHEx-BE: the heat pump is directly connected to the building thermal load;
(c) TS-BE: the building thermal load is met by the thermal storage.

The water content of the primary loop corresponds to a typical volume of a single-family heating system, namely around 45 L. This can also test the effects of the water loop dynamics in a real system. The volume also includes the expansion vessel and the inertial storage required by the AHP (about 7 L/kW).

The BE consists of an oversized outdoor heat exchanger controlled by a three-way mixing valve. The emulator produces a fluid enthalpy variation equal to the thermal power evaluated by the dynamic simulation of the simulated building (see Section 3). The outdoor air–water heat exchanger has an oversized nominal capacity with respect to the rest of HiL apparatus to ensure the heat transfer when the outdoor air and the secondary loop have a close temperature.

The thermal storage (520 L) is connected to the user loop and can be used to provide thermal energy to the BE in replacement of the AHP. The system does not exploit the bulk fluid that is intended to be used for domestic hot water purposes. The energy transfer occurs through an oversized internal coil. The electrical resistance (4.8 kW) is the standard TS electrical back-up, however, it can also be controlled to deliver a given profile of thermal power, thus acting as an emulator of other generation technologies.

Table 1. Components of the BE-IES "Hardware-in-the-Loop" Apparatus.

Component	Role	Description
Air-to-water heat pump (AHP)	Generation	Nominal heating capacity [1]: 5.8 kW Nominal COP [1]: 4.53 Nominal $\eta^{II} = 0.41$ Variable-speed compressor: 20–72 Hz
Pump A		Nominal flowrate: 980 L/h
Main heat exchanger (mHEx)	Interface between generation and user sections	Nominal capacity: 31 kW ($\overline{\Delta T}_{f,a} = 12$ K)
Thermal storage (TS)		Volume: 0.52 m^3. Nominal coil capacity: 20 kW ($\overline{\Delta T}_{f,b} = 35$ K)Insulation coefficient: 1.5 W/K Back-up electrical resistance: 4.2 kW
Weather station		Measured quantities: dry bulb temperature, relative humidity, air pressure, global solar irradiance on a horizontal surface, infrared sky radiation, wind speed and direction
AHP-TS switching valves	Thermal dispatchment	
Outdoor air-water heat exchanger (OutHEx)	Building emulator	Nominal capacity: 20 kW ($\overline{\Delta T}_{f,a} = 12$ K)
Three-way mixing valve	Building emulator	Nominal flow coefficient: $K_{Vs} = 1.6$ m^3/h $(0-10\text{ V})$
Pump B		Variable-speed pump $(0-10\text{ V})$. Max flow rate: 2.8 m^3/h; Max head: 5 m
Ducts		1′ $\frac{1}{4}$ Polypropylene random copolymer (PP-R)

[1] Rating conditions in accordance with EN 14511-2:2018: Water supply/return: 35 °C/30 °C; Ambient air conditions: 7 °C DBT/6 °C WBT.

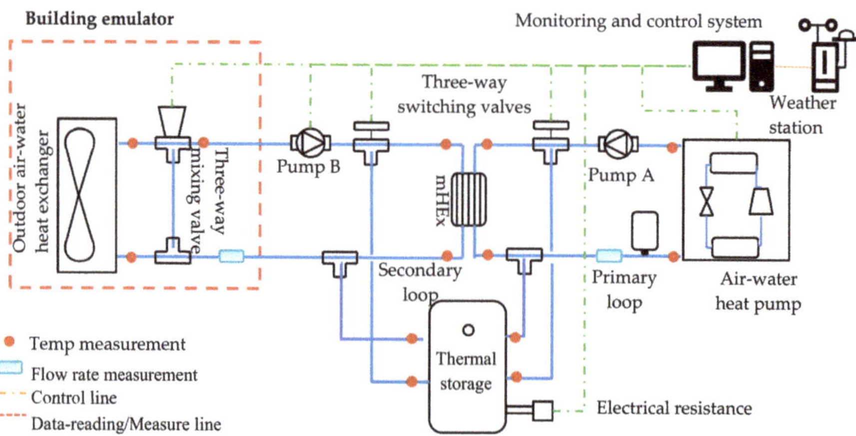

Figure 2. Layout and components of the BE-IES "hardware-in-the-loop" apparatus.

2.2. Measurement System

The HiL apparatus is equipped with industrial measurement devices with the typical performances of such technologies. The choice of not using accurate laboratory equipment is motivated by the will to develop control strategies that can be performed by commercial HVAC devices and their typical monitoring, control, and actuator systems.

The temperature is read by eleven Negative Temperature Coefficient (NTC) thermistors, located as in Figure 2. NTC thermistors were chosen because of their high accuracy in the operational temperature range of the HiL apparatus, together with their ease of installation with respect to other types of sensors (e.g., thermocouples). All the NTC were calibrated before being installed in the apparatus. The accuracy of each thermistor in the −5 ÷ 50 °C range was evaluated as equal to ±0.20 K (confidence level of 95.4%, i.e., 2σ). The corresponding accuracy on a temperature drop measure was ±0.3 K. For fluid measurements, the NTC thermistors were stuck inside a copper probe located in approximately the middle of the flow.

The flow rate in the two loops was measured through two industrial Kármán vortex flowmeters, located as in Figure 2. The calibration procedure revealed an overall accuracy in the 5 ÷ 25 L/m range of ±0.0225 L/m (confidence level of 95.4%, i.e., 2σ).

The resulting accuracy on the enthalpy drop measures, namely the thermal power, was evaluated through the error propagation theory. For instance, the accuracy in proximity of the nominal AHP flow rate (980 L/h) and temperature drop (5 K) of the AHP is about 320 W, which corresponds to a relative error of about 5.52% with respect to the nominal capacity (5.800 kW).

The weather station is located on the roof of the laboratory building. The accuracy of each sensor is shown in Table 2.

Table 2. Accuracy of the Weather Station Instruments.

Sensor	Measured Quantity	Accuracy
Thermistor (PT 100)	Dry bulb temperature (°C)	±0.15 K ± 0.1%
Metal oxide's electrical capacity	Relative humidity (0–1)	±0.015 ± 1.5%
Piezoresistive pressure sensors	Air pressure (Pa)	±50 Pa
Ultrasonic anemometer	Direction and speed	±1° (direction) ±0.2 m/s ± 2% readings (speed)
Thermopile Pyranometer	Global solar irradiance $\left(W/m^2\right)$ in the wavelength range 0.30 ÷ 2.80 µm	±10%
Thermopile Pyrgeometer	Infrared sky radiation $\left(W/m^2\right)$ in the wavelength range 5.5 ÷ 45 µm	±10%

2.3. Monitoring and Control System: Operational Strategy of the HiL Apparatus

All the measurements are collected by a desktop computer through some in-house developed LabVIEW® virtual instruments (VIs). The same software is used to run the dynamic simulation of the building and to control the HiL apparatus (i.e., building emulator, valves, and AHP), according to the operational scheme already presented in Figure 1 and hereafter discussed in detail. The dynamic simulation of the building is based on an in-house validated algorithm written in MATLAB® language, corresponding to the thermal network model described in Section 3.

According to the chosen monitoring and simulation time step, the M&C system collects the weather data, the fluid temperature and the flow rate in the various HiL sections, together with the electrical energy input of AHP and ancillary systems. Then, it uses the actual weather data, the supply temperature to the BE, and the current thermal state of the simulated building to evaluate the performance of the simulated heat emission units (see Section 3). The latter heat output and the actual climate data are thus used to simulate the thermal evolution of the building for the next time step. At the same time, the three-way mixing valve within the BE is controlled though a Proportional-Integral-Derivative (PID) controller to deliver the simulated thermal performance at the mHEx (or in the TS). The procedure is repeated at each time step and emulates the dynamic operation of an actual AHP-TS system connected to the simulated building. Depending on the chosen control strategy, the AHP can be switched off or used to heat-up the thermal storage by controlling the AHP-TS switch valve.

3. Dynamic Simulation Model of the Building and Heat Emission System

The experimental analysis based on the HiL system requires an accurate energy model of the building to run the BE. Figure 3 shows the main heat fluxes affecting the energy balance of a generic building or thermal zone that has been included in the building dynamic energy model, namely: heat transfer through opaque and glazed surfaces, ventilation losses, heat gains related to solar radiation, occupants, and electrical appliances. For the sake of clarity, in Figure 3 we do not display the energy stored in building masses, such as external and internal walls. However, these quantities and their effects in the thermal evolution of the building are included in the dynamic energy model.

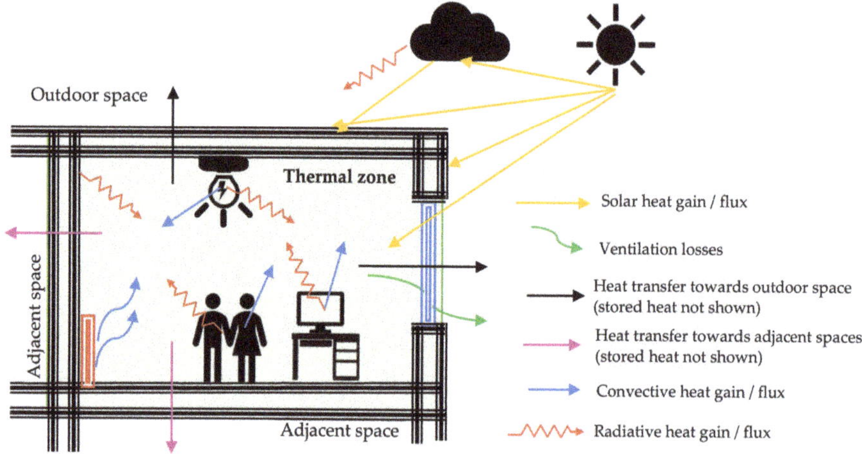

Figure 3. Illustrative scheme of the main energy/heat fluxes affecting the building energy balance (stored energy contributions are not shown).

In this work, we used a classical thermal network approach, widely used at both research and professional levels (see, for instance, EN ISO 52016-1:2017 [25]). The building elements are modeled as electrical nodes connected to each other by equivalent electrical resistances, capacities, time-dependent currents and voltage generators. A summary list of considered phenomena and the corresponding electrical elements can be found in Section 2.1 of [26], including all the info on the thermal-network model and its validation. The building dynamic model corresponds to a $(N \times N)$ linear set of equations that can be quickly solved at each timestep. N is the number of thermal nodes considered in the model, including the one of indoor air.

The evolution of nodes' temperature is a function of climate data (e.g., outdoor air temperature, solar irradiance, sky temperature), heat gains (e.g., people, electrical appliances, and lighting), and heat exchanged by the system terminals. All these elements are modeled as voltage or current generators: their source value is updated at each timestep, according to a given profile (e.g., occupants' presence) or measured quantities in the HiL apparatus (e.g., climate data and supply temperature to the heat emission systems). The heat gain delivered by the emission system can be evaluated through a few additional linear equations to be added to the overall set representing the building. In this work, we refer to fan coil units (FCUs) as heat terminal units. The heat transfer performance is modeled through the classical methods of heat exchangers theory, namely

$$\begin{cases} \dot{Q}_{FCU} = \dot{m}_f c_f \left(T_{f,in} - T_{f,out}\right) \\ \dot{Q}_{FCU} = \dot{m}_{a,vel} c_a \left(T_{a,out} - T_{a,in}\right) \\ \dot{Q}_{FCU} = (UA)_{vel} \overline{\Delta T}_{f,a} \end{cases} \quad (1)$$

The set of Equation (1) allows the evaluation of $\dot{Q}_{FCU}$, $T_{a,out}$, $T_{f,out}$ as a function of the inlet air temperature, $T_{a,in}$, inlet fluid temperature, $T_{f,in}$, air flow rate, $\dot{m}_{a,vel}$, water flow rate, $\dot{m}_f$, and overall heat transfer coefficient $(UA)_{tot,vel}$, evaluated from FCU manufacturer's datasheet. The subscript *vel* indicates that different values of the coefficient are used according to the fan speed (generally, high, medium, and low).

The heating performance of the fan coils and the corresponding heat gain for the indoor air node are evaluated at each timestep. If $\overline{\Delta T}_{f,a}$ is approximated with the difference between the arithmetic means of water and air temperature, the set of Equation (1) is linear and it can be added and solved in conjunction with the overall algebraic model of the building. Otherwise, classical numerical techniques for non-linear set of equations can be used. Shortly, the heat emission system is modelled as two additional thermal nodes, $T_{f,in}$ and $T_{f,out}$, in the overall building thermal network. $T_{f,in}$ is set equal to the measured temperature of the fluid at the inlet section of the BE; $T_{f,out}$ is an unknown of the problem to be calculated at each timestep. Its value is then applied as the set point for the three-way mixing vale in the BE.

4. Description of the Case Study

The above-described HiL apparatus was used to emulate the seasonal winter performance of a commercial AHP acting as a heat generator in the selected building: a typical single-family house.

4.1. Reference Boilding and Heat Emission System

The case study is a typical apartment located at the top floor of a building (see Figure 4), situated in Pisa, Italy. Having a single thermostat set at 20 °C, according to accepted definitions (see, for instance [27]), the whole apartment can be modeled as a single thermal zone, with a total floor area and volume equal to 84 m^2 and 227 m^3, respectively. The thermal zone has four external walls with double pane windows (N, E, S, and roof); the western wall and the floor are adjacent to similar apartments. The stratigraphy of walls is typical of Italian structures in the 1980s: the external walls are made of two layers of bricks, concrete and plaster for a total width of about 50 cm; the wall at West is similar to the external ones, but the width is reduced to about 30 cm; the internal walls are light elements made of narrow bricks (about 10 cm); the external roof is horizontal and entirely practicable, the supporting structures are made of 30-cm slabs reinforced with concrete beams. The roof is not considered to be covered by snow for a significant amount of time (with consequent effects on its heat exchange properties during wintertime), since it very rarely snows in Pisa. The internal walls between rooms are included in the analysis to account for their thermal inertia. The overall transmission coefficients of walls and windows are reported in Table 3. Air change rate has been set as equal to 0.5 volumes per hour. Occupants, lighting, and electrical appliances are included in the building energy balance as heat gains. The corresponding hourly profiles are shown in Figure 5.

This design thermal load of the case study (about 5.7 kW) is coherent with the thermal capacity of tested AHP. Considering the number of rooms, the assumed seven equal fan coils located within the thermal zone. The nominal data depending on fan speed are provided by manufacturer's datasheet (see Table 4) according to the rating conditions of EN 1397:2015 [28].

Table 3. Geometry and Thermal Characteristics of the Envelope.

Wall	Surface m^2	Overall Transmittance $W/(m^2K)$	Weight kg/m^2
Roof	84	0.6	485
Northern	19	0.6	345
(glazed)	(2 × 3.36)	2.8	
Eastern	32	0.6	345
(glazed)	(2 × 1.96)	2.8	
Southern	14	0.6	345
(glazed)	(4.2)	2.8	
Western	38	1.5	128
Floor	84	1.2	339
Internal walls	79	1.5	120

Total external (opaque–glazed–total): 201 – 7.5 – 209 m^2
Surface-volume ratio: 0.91
Overall weight of wall per floor area: 1232 kg/m^2

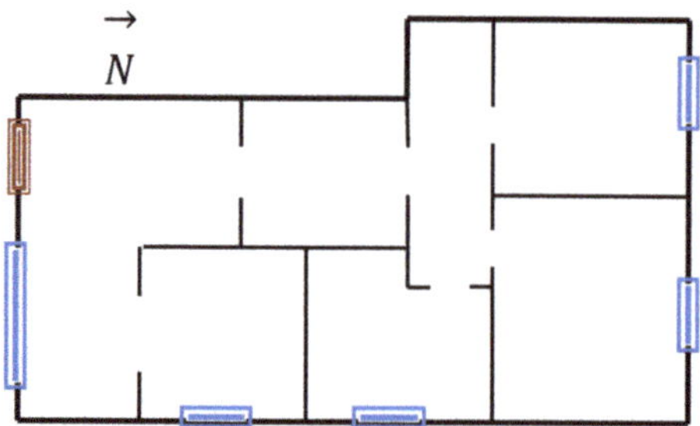

Figure 4. Scheme of the apartment (blue boxes are windows, while the brown box is the main door).

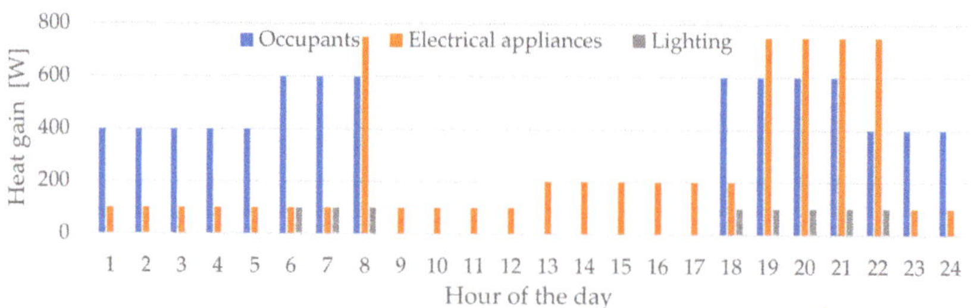

Figure 5. Daily profiles of internal heat gains for the apartment.

Table 4. Fan coil unit (FCU) Manufacturer's Datasheet [1].

	Max	Med	Min
Thermal power, kW	1.00	0.73	0.52
Water flow rate, L/h	174	126	92
Air flow rate, m^3/h	180	120	80

[1] Rating conditions in accordance with EN 1397:2015 [28]: Water supply/return: 45 °C/40 °C; Ambient air conditions: 20 °C DBT.

4.2. Test Description and Objectives

The experimental campaign was performed from November 2019 to February 2020 during the heating period. The main objective was the measurement of the seasonal performances of the AHP and of the whole HVAC-building system in real operational conditions, thus subjected to various on/off and modulation cycles because of weather and heat gains evolution, building thermal inertia, water loop and AHP generator dynamic characteristics.

Different simplified, but realistic, AHP control modes were tested. As previously mentioned, a single-zone thermostat controlled the on/off signal to the AHP with an indoor setpoint of 20 ± 1 °C. Three operational modes of the fan coils were tested in order to evaluate their effect on the system performance: MOD#1 corresponds to a setpoint value of the supply temperature equal to 40 °C and MED fan coil speed; MOD#2 uses the same supply temperature but the fan coil speed was reduced to MIN; MOD#3 has a supply temperature set equal to 35 °C and MIN as fan coil speed. The thermal storage was not used in this test campaign as it is not usually present in apartments of multi-apartment buildings with autonomous heating system.

Two seasonal coefficients of performance were chosen as performance indexes to compare the three operational modes: $SCOP_{AHP}$ is defined as the ratio of the thermal energy output and electrical energy input at the AHP devices over a given operational period (see Equation (2)); $SCOP_{sys}$ is defined as the ratio of the thermal energy delivered to the building emulator (i.e., simulated fan coils) and electrical energy input at the AHP devices over a given operational period (see Equation (3)).

$$SCOP_{AHP} = \frac{\int_{\tau} \dot{m}_{AHP,f} c_f (T_{AHP,out} - T_{AHP,in}) d\tau}{\int_{\tau} \dot{W}_{AHP,in} d\tau} \quad (2)$$

$$SCOP_{sys} = \frac{\int_{\tau} \dot{m}_{FCU,f} c_f (T_{FCU,in} - T_{FCU,out}) d\tau}{\int_{\iota} \dot{W}_{HP,in} d\tau} \quad (3)$$

The two performance indexes allow a separate analysis of the heat generator and of the overall building-HVAC system. $SCOP_{sys}$ only refers to the useful heat delivered to the building thermal zone, thus it can be used to evaluate the relevance of the thermal losses associated to the distribution pipework and to the control of the emulated heating system.

5. Results

In this section, we discuss how the system operates under the three control strategies during the whole test period. To better analyze the dynamics of operative quantities, Figures 6–8 show some typical days under MOD#1, MOD#2, MOD#3 in a typical mild (left subfigures) and cold days (right subfigures). The upper subfigures show the profile of the AHP and FCU thermal outputs; the two profiles are zero when the heating system is switched off. The lower subfigures show the evolution of some relevant temperatures, namely, the AHP supply, the AHP return, indoor air, and external air. We note that the AHP manages to maintain the thermal comfort conditions within the apartment (i.e., $T_i = 20 \pm 1$ °C) in all the periods, independently from the control mode.

In this work, we compare the experimental performances of the AHP with the ones evaluated through the state-of-the-art methodology for calculation of heat pump energy requirements and efficiency presented in technical standards EN 15316-4-2:2017 [22] and EN 14825:2018 [24]. The method is based on manufacturers' datasheets and represents the main reference for energy dynamic simulations, energy audits, and cost-benefit analysis of heat generation alternatives. The comparison is thus aimed at evaluating the effectiveness of the standard methodology to properly account for the penalization effects due to the operation dynamics. On the other hand, quantitative comparisons with similar works, in which a modulating air-to-water heat pump is monitored when coupled to a HiL system [10] or to an existing building [29,30], or it is fully simulated under dynamic building-coupled conditions [31], are somehow misleading, since energy performances are very heterogeneous, being strongly influenced by climatic conditions and building and heat pump characteristics. However, in these cases, a meaningful parameter for comparison purposes is the penalization factor due to partial loads (see Appendix A). In this regard, the measured performance reduction goes from about 10% [32–34] to over 30% [35], depending on the frequency of the on–off cycles.

5.1. Analysis of MOD#1 Operational Strategy: Energy Losses Due to On-Off Cycles

When MOD#1 applies ($Vel = MED$; $T^{*}_{out,AHP} = 40$ °C), the profiles are characterized by continuous on–off cycles of about 40 min (see Figure 6). The average duration of each on period, approx. 20 min, is not sufficient to achieve a steady state condition of the water loop, thus the supply and return temperatures do not reach the setpoint values and continuously fluctuate. However, the FCU thermal output during this transient period is enough to heat up the indoor air until the switching off conditions (i.e., 21 °C); the heat pump was thus deactivated without any modulation of the compressor speed. Figure 9 shows the distribution of compressor frequency: when MOD#1 applies, the frequency was null or about 49 Hz for more than 80% of the time. Figure 10 shows the distribution of the duration of the ON periods: when MOD#1 applies, ON periods of 15 – 30 min covered more than 70% of the total operative time.

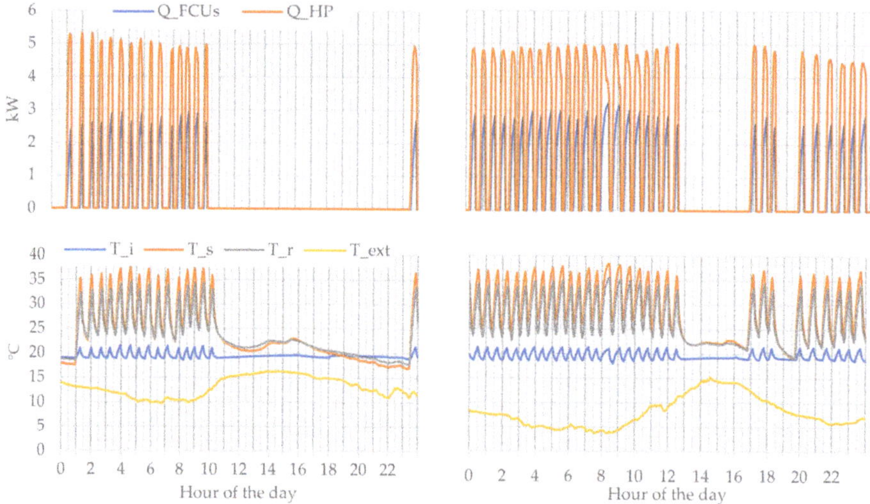

Figure 6. Typical mild (**left**) and cold (**right**) days under MOD#1 (23-Dec., 5-Jan.).

Heat pump efficiency is affected by ON–OFF cycles: the experimental $SCOP_{AHP}$ under MOD#1 is about 32% lower than the one evaluated according to manufacturer's datasheet (3.07 vs. 4.53) at the same average outdoor temperature (9 °C), water supply temperature (34 °C), and integral-averaged capacity ratio (0.26), defined as the ratio of the thermal energy output and the maximum energy deliverable at nominal heating capacity. The deviation between the two performance indexes reflects the reduced accuracy of standardized methodologies and manufacturers' datasheets when the operational regime strongly fluctuates.

We also note that the generator has always provided a thermal power close to the nominal capacity (about 5 kW) and almost twice the one exchanged by the FCUs (see Table 5 and Figure 11). The deviation between the two power profiles represents the internal energy variation in the heating loop that is increasing its temperature. This additional energy does not result in useful heat for the building and it will be lost during the following OFF period when the water loop cools down. The energy losses associated with the just-described behavior are quantified by $SCOP_{sys}$ which is about 35% lower than while $SCOP_{AHP}$ (2.00 vs. 3.07).

Finally, we note that the intermittent work regime occurred for both mild and cold outdoor temperatures (see Figure 5): a colder climate only increased the number of ON–OFF cycles as the FCU thermal output has been always greater than building heating demand.

5.2. Analysis of MOD#2 Operational Strategy: Effects of Fan Coils' Speed

MOD#2 ($Vel = LOW$; $T^*_{s,AHP} = 40$ °C) shows a little more regular operation than of MOD#1 (see Figure 7), but the intermittent regime was maintained. The typical duration of ON periods almost doubles from 20 to 38 min (see Figure 9), but the number of ON–OFF cycles is still relevant for both cold and mild days. We note that the heating system reached a steady state condition in a few hours during night-time due to a reduced output of the heat emission system (see Table 5). The longer period required to reach the switching off condition (i.e., 21 °C) was sufficient for the heat pump to modulate its compressor.

However, the actual heat pump performance was 35% lower than the one evaluated according to the manufacturer's datasheet (2.95 vs. 4.56) at the same average outdoor temperature (9 °C), water supply temperature (36 °C), and integral-averaged capacity ratio (0.31). The deviation between $SCOP_{AHP}$ and $SCOP_{sys}$ (i.e., 2.95 vs. 2.02) confirms the relevance of the thermal losses due to the continuous variation in water loop temperature and the low significance of manufacturers' datasheets in intermittent work regimes.

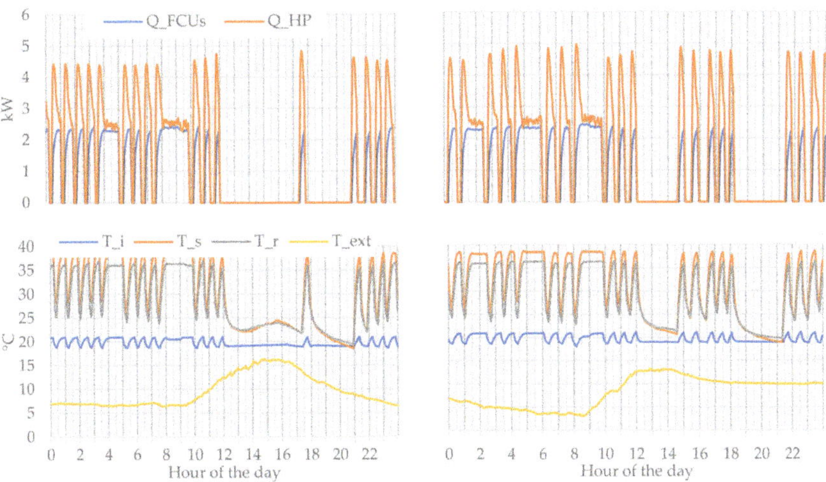

Figure 7. Typical mild (**left**) and cold (**right**) days under MOD#2 (22-Jan., 19-Jan.).

5.3. Analysis of MOD#3 Operational Strategy: Effects of Supply Temperature

MOD#3 ($Vel = LOW$; $T^*_{s,AHP} = 35$ °C) shows an improved behaviour of the system. The heat pump operation is characterized by a few constants ON periods, mainly occurring at night-time (see Figure 8). The average duration of ON periods under MOD#3 is shown in Figure 10: we note long ON periods of some hours. The average AHP thermal output during ON periods decreases from 4.00 kW in MOD#1 to 1.39 kW in MOD#3, which is a more similar value to the FCU thermal output (1.17 kW). Shortly, under MOD#3, the HVAC system reaches a steady state condition and a thermal balance with the building heating demand. The losses of the heating loop (0.12 kW) were reduced with respect to MOD#1 (0.55 kW) and MOD#2 (0.59 kW) due to the absence of the continuous alternation of additional heating-up requirements and deactivation losses. Under MOD#3, the compressor has enough time to start modulating its speed, mainly working with a compressor frequency in the 20–30 Hz range (see Figure 9). These work conditions decrease the inefficiencies due to ON–OFF cycles; consequently, $SCOP_{AHP}$ increases to 3.03 and $SCOP_{sys}$ increases to 2.56. However, we note that the actual heat pump performance was 32% lower than the one evaluated according to manufacturer's datasheet (3.03 vs. 3.52) at the same average outdoor temperature (11 °C), water supply temperature (33 °C), and integral-averaged capacity ratio (0.13).

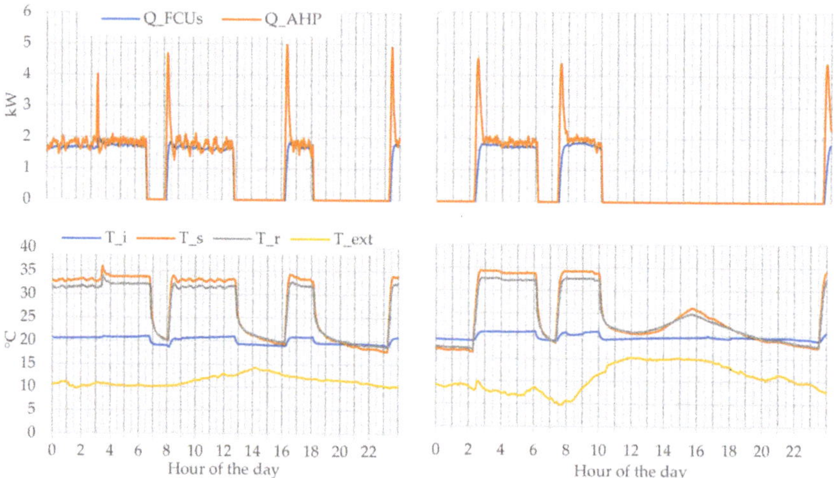

Figure 8. Typical mild (**left**) and colder (**right**) days under MOD#3 (25-Jan., 5-Feb.).

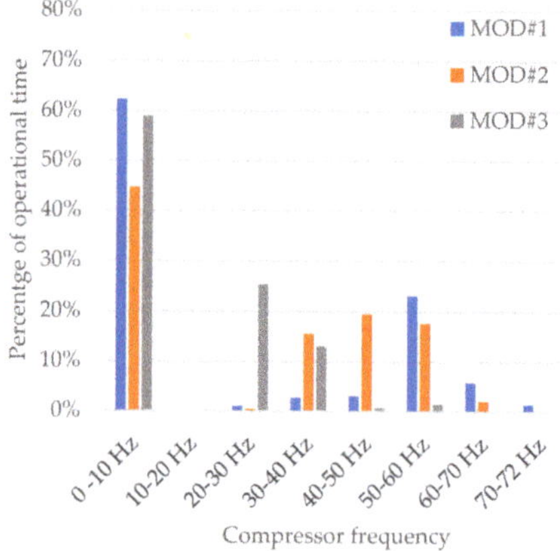

Figure 9. Compressor frequency distribution as percentage of time under each operational mode.

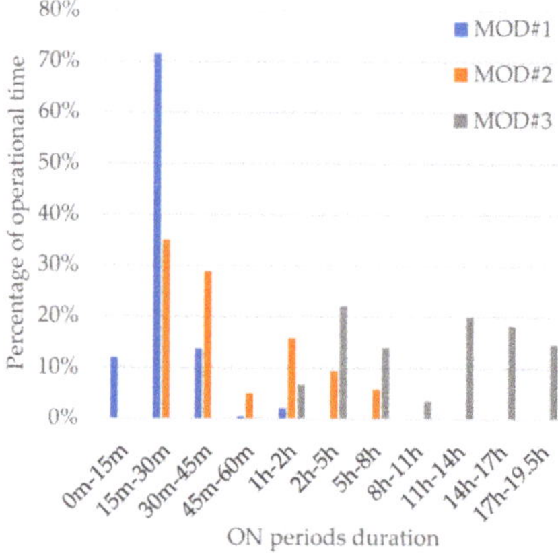

Figure 10. Distribution of the ON periods duration under each operational mode.

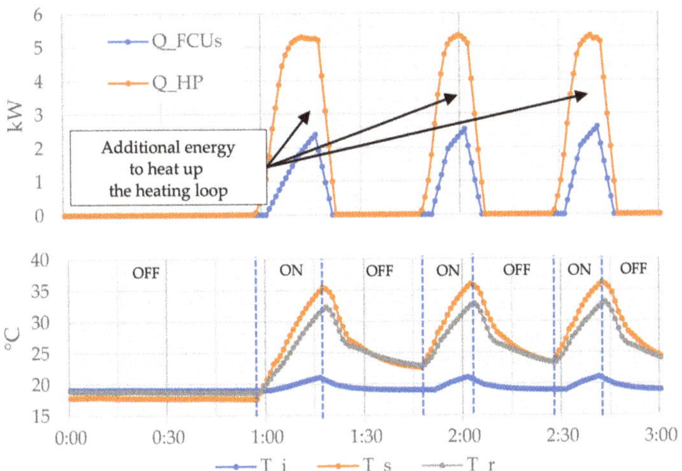

Figure 11. Detail of ON–OFF cycles under MOD#1.

Table 5. Summary of AHP and Overall System Performances.

	MOD#1 *Vel = MED* $T^*_{s,AHP}$ = 40 °C	**MOD#2** *Vel = LOW* $T^*_{s,AHP}$ = 40 °C	**MOD#3** *Vel = LOW* $T^*_{s,AHP}$ = 35 °C
$SCOP_{AHP}$	3.07	2.95	3.03
$SCOP_{sys}$	2.00	2.02	2.56
Average heat pump output during ON periods–kW	4.14	3.34	1.39
Average fan coils output during ON periods–kW	2.28	1.90	1.17
Average supply temperature during ON periods–°	33.7	35.6	33.0
Average outdoor temperature during ON periods–°C	9.0	8.7	11.0
Integral-averaged AHP capacity ratio during ON periods	0.69	0.57	0.22
Integral-averaged AHP capacity ratioover the total test period	0.26	0.31	0.13
Integral-average losses over the total test period–kW	0.55	0.59	12
$SCOP_{AHP}$ (according to manufacturer's datasheet and reference technical standards)	4.53.	4.56	3.52

In conclusion, for all the three control strategies, the experimental data showed a considerable reduction (30–45%) of the coefficients of performance with respect to the ones based on manufacturer's datasheet and technical standards.

Moreover, in any MOD, the actual performances of the heat pump were about 3. The latter value is lower than the expectations considering the mild climate of Pisa. With reference to the Italian residential energy market, we recall that the break-even seasonal performance required for an electrically driven heat pump to be economically advantageous with respect to a gas boiler is in the range 2.2–2.5, depending on actual prices of natural gas, electricity, and boiler efficiency.

6. Conclusions

This paper presents the results the HiL extensive experimental campaign performed at BE-IES laboratory of DESTEC (University of Pisa). We measured the actual performance of an air heat pump system used in a typical residential building, during the winter season, according to three different operational modes. We also compared the experimental performances with the corresponding ones evaluated through the state-of-the-art methodology presented in current technical standards (i.e., [22,24]). The emulated building consists of an apartment located at the top floor of a multi-apartment building in Pisa. The simulation/emulation periods went from November 2019 till February 2020: about 740 h of data were collected every minute. The results pointed out three main findings:

- The typical thermal design load results in an overestimated value of FCU capacity and the thermal zone thermostats continuously sent an ON–OFF signal to the heat generator. The heating system never reached a steady state condition under MOD#1, with notable losses due to the continuous heating-up and cool-down phases occurring twice per hour. The reduction in supply temperature from 40 to 35 °C has been the key to ensure smoother operations. The average duration of the ON periods increased from 20 min in MOD#1 to about 10 h in MOD#3. The fan speed reduction resulted in a less effective strategy to reduce the ON–OFF cycles;
- The experimental performance of the AHP device in MOD#1 and MOD#2, $SCOP_{AHP}$, was about 35% lower than the one evaluated through current state-of-the-art methodologies, based on manufacturer's datasheet and empirical correlations for accounting partial-load conditions (3.07 vs. 4.53 and 2.95 vs. 4.56). The deviation between experimental and simulated results reduced to 14% when the system operated in smoother regimes, namely in MOD#3 (3.03 vs. 3.52). The actual $SCOP_{AHP}$ value was about 3 and this is quite above the threshold value for an electrically driven heat pump to be convenient with respect to a gas boiler (minimum *SCOP* about 2.2–2.5);
- The performance of the overall building-HVAC system is notably reduced by the transient regime of the heating system. $SCOP_{sys}$ is about 50% of $SCOP_{AHP}$ in MOD#1 and MOD#2 and about 16% in MOD#3. At the end, the seasonal performance of the overall HVAC-building system, $SCOP_{sys}$, was equal to 2.00, 2.02, 2.56 which is lower than the values expected in a mild climate such as the one of Pisa. The main losses are associated with the continuous heating-up and deactivation sequences: the delivered energy to heat up the water after an off period was lost at the next off period. In such conditions, the thermal losses of the water loop are relevant with respect to the heat provided by the AHP (32% in MOD#1 and MOD#2 and 15% in MOD#3).

The analysis here presented demonstrates the potential of the HiL approach to properly estimate energy consumptions of and available improvements to HVAC systems under real operative conditions. The utilization of standardized methodologies, even if based on manufacturers' datasheets, can lead to significative deviations from actual performances due to the difficulty of capturing the transient performances of the devices and the interaction between building and HVAC dynamics.

The HiL approach is also promising for developing advanced operational strategies. In the presented case study, the use of a weather-base control for the supply temperature is expected to be advantageous. The use of a thermal storage can be beneficial as well: it can be used to store heat during the hottest hours of the day, when the building does not require heating and the outdoor temperatures are favorable. Additionally, ON–OFF losses will be reduced as the heat pump will experience a constant operation during the "charging" phases of the storage. In other words, the stored heat will be delivered to the building with a peak-shaving effect. The latter improvement can also be achieved with smaller storage volumes and will be the goal of another experimental campaign at the BE-IES laboratory.

Author Contributions: Conceptualization, C.B., P.C., A.F. and D.T.; methodology, C.B., P.C., A.F. and D.T.; software, P.C.; formal analysis, C.B., P.C., A.F. and D.T.; data curation, P.C.; writing—Original draft preparation, P.C.; writing—Review and editing, C.B., P.C., A.F. and D.T. All authors have read and agreed to the published version of the manuscript.

Funding: The authors gratefully acknowledge the financial support of the Italian Ministry of Education, University and Research (MIUR), in the framework of the Research Project of Relevant National Interest (PRIN) "The energy FLEXibility of enhanced HEAT pumps for the next generation of sustainable buildings (FLEXHEAT)" (PRIN 2017, Sector PE8, Line A, Grant n. 33).

Acknowledgments: We acknowledge the other colleagues, technical staff and students of DESTEC department of the University of Pisa, who contributed to the development of the BE-IES laboratory and to the experimental campaign, with special thanks to Davide Della Vista, Roberto Manetti, Riccardo Dinelli, Alessandro Conforti, and Niccolò Malfanti.

Conflicts of Interest: The authors declare no conflict of interest.

Nomenclature

Acronyms	
AHP	Air-to-water heat pump
BE-IES	Building Emulator and Integrated Energy Systems laboratory
BE	Building Emulator
DBT	Dry-bulb temperature, °C or K
FCU	Fan coil unit
HiL	Hardware-in-the-loop system
HP	Heat Pump
HVAC	Heating Ventilation and Air Conditioning
M&C	Monitoring and control system
mHEx	main Heat Exchanger
NTC	Negative Temperature Coefficient thermistors
PID	Proportional Integral Derivative
SCOP	Seasonal Coefficient of Performance
TS	Thermal storage
VI	LabVIEW® virtual instruments
WBT	Wet-bulb temperature, °C or K
Symbols	
C_d	degradation coefficient
CR	capacity ratio
c	specific heat, J/(kg K)
$\dot{Q}$	thermal power, W
$SCOP$	mean or Seasonal Coefficient of Performance
T	Temperature, °C or K
UA	*heat transmittance-surface product,W/K*
$\dot{W}$	electrical power, W
$\overline{\Delta T}$	*Mean temperature difference,°C orK*
η^{II}	Second law efficiency
τ	time, s
Subscripts	
a	air
ext	external air
f	fluid
i	indoor air
in	inlet or input
out	outlet or output
r	return
s	supply
vel	fan coil speed
Superscripts	
*	setpoint value

Appendix A. AHP Manufacturer's Datasheet and State-Of-The-Art Methodology Based on Second-Law Efficiency

The most common methodology at both a professional and research level to assess the heat pump performances is based on the interpolation of the so-called second-law or exergy efficiency, η^{II}. The procedure is based on the manufacturers' experimental data provided in accordance with EN 14511-2:2018 and EN 14825:2018 (see, for example, Table A1). The data on these datasheets are generally marked with the subscript DC that indicates a quantity evaluated at maximum compressor speed. The second-law efficiency η^{II}_{DC} is defined as in Equation (A1) and it is calculated for each COP_{DC} value provided by the manufacturer

$$\eta^{II}_{DC} = COP_{DC}\frac{T_{out,HP,DC} - T_{a,in,HP,DC}}{T_{out,HP,DC} + 273.15} \quad \text{(A1)}$$

Then, η^{II}_{DC} is interpolated according to the actual supply and outdoor temperature $T_{out,HP}$ and $T_{a,in,HP}$. We refer to the result of the interpolation procedure as η^{II}. Similarly, the maximum power output of the device at the actual temperature conditions, $\dot{Q}_{max}\left(T_{out,HP}; T_{a,in,HP}\right)$, is evaluated through interpolation of $\dot{Q}_{out,DC}$. The capacity ratio of the device is evaluated according to the average actual thermal output, $\dot{Q}$, over the considered time (off periods included), namely

$$CR = \frac{\dot{Q}_{out}}{\dot{Q}_{max}\left(T_{out,HP}; T_{a,in,HP}\right)} \tag{A2}$$

According to the above-mentioned technical standards, it is possible to evaluate a penalization factor, f_{CR}, as a function of CR and its capacity control system. For variable capacity units, Equation (A3) applies

$$f_{CR}(CR) = f(x) = \begin{cases} 1, & CR \geq CR_{min} \\ \frac{CR}{CR \times C_d + (1 - C_d)}, & CR < CR_{min} \end{cases} \tag{A3}$$

The degradation coefficient C_d and the minimum control level, CR_{min}, can be provided by the manufacturer or read in EN 14825:2018. Finally, the actual COP and energy input is evaluated as

$$COP\left(T_{out,HP}; T_{a,in,HP}; CR\right) = \frac{T_{out,HP} + 273.15}{T_{out,HP} - T_{a,in,HP}} \times \eta^{II}\left(T_{out,HP}; T_{a,in,HP}\right) \times \ f_{CR}(CR) \tag{A4}$$

$$\dot{W}_{in} = \frac{\dot{Q}_{out}}{COP\left(T_{out,HP}; T_{a,in,HP}; CR\right)} \tag{A5}$$

The procedure is generally applied on an hourly or monthly time step.

Table A1. Manufacturer's Datasheet of the Considered Air-to-Water Heat Pump [1]. Grey Boxes Refer to Nominal Data.

$T_{out,HP}$ °C	25		30		35		40		45		50		55	
$T_{a,in}$ °C	$\dot{Q}_{out,DC}$ [kW]	$\dot{W}_{in}$ [kW]	$\dot{Q}_{out,DC}$ [kW]	$\dot{W}_{in,DC}$ [kW]	$\dot{Q}_{out,DC}$ [kW]	$\dot{W}_{in,DC}$ [kW]	$\dot{Q}_{out,DC}$ [kW]	$\dot{W}_{in,DC}$ [kW]	$\dot{Q}_{out,DC}$ [kW]	$\dot{W}_{in,DC}$ [kW]	$\dot{Q}_{out,DC}$ [kW]	$\dot{W}_{in,DC}$ [kW]	$\dot{Q}_{out,DC}$ [kW]	$\dot{W}_{in,DC}$ [kW]
−20	5.24	1.83	5.01	2.07	4.78	2.3	4.55	2.54	-	-	-	-	-	-
−15	5.67	1.83	5.44	2.07	5.21	2.3	4.98	2.54	4.75	2.78	-	-	-	-
−10	5.96	1.8	5.8	2.04	5.51	2.27	5.49	2.5	5.33	2.73	5.17	2.96	5.01	3.19
−7	6.14	1.79	6.02	2.02	5.91	2.25	5.79	2.47	5.67	2.7	5.56	2.93	5.44	3.16
−2	6.44	1.56	6.31	1.77	6.18	1.98	6.06	2.19	5.93	2.4	5.8	2.6	5.67	2.81
2	6.51	1.29	6.37	1.48	6.24	1.66	6.1	1.85	5.97	2.03	5.83	2.22	5.7	2.4
7	6.3	1.01	6.05	1.15	5.8	1.28	5.55	1.42	5.3	1.55	5.05	1.69	4.8	1.82
10	6.48	1.03	6.25	1.17	6.01	1.3	5.77	1.44	5.53	1.57	5.29	1.71	5.05	1.84
15	6.79	1.07	6.57	1.2	6.35	1.34	6.13	1.47	5.92	1.61	5.7	1.74	5.48	1.88
20	7.1	1.1	6.9	1.24	6.7	1.37	6.5	1.51	6.3	1.64	6.1	1.78	5.9	1.91

[1] Rating conditions in accordance with EN 14511-2:2018. Heating capacity is according to Eurovent rating standard RS-6/C/001-2011 and valid for heated water range $\Delta T = 3 \sim 8$ °C.

References

1. Fischer, D.; Madani, H. On heat pumps in smart grids: A review. *Renew. Sustain. Energy Rev.* **2017**, *70*, 342–357. [CrossRef]
2. Franco, A.; Salza, P. Strategies for optimal penetration of intermittent renewables in complex energy systems based on techno-operational objectives. *Renew. Energy* **2011**, *36*, 743–753. [CrossRef]
3. Testi, D.; Urbanucci, L.; Giola, C.; Schito, E.; Conti, P. Stochastic optimal integration of decentralized heat pumps in a smart thermal and electric micro-grid. *Energy Convers. Manag.* **2020**, *210*, 112734. [CrossRef]
4. Conti, P.; Lutzemberger, G.; Schito, E.; Poli, D.; Testi, D. Multi-Objective Optimization of Off-Grid Hybrid Renewable Energy Systems in Buildings with Prior Design-Variable Screening. *Energies* **2019**, *12*, 3026. [CrossRef]
5. Testi, D.; Conti, P.; Schito, E.; Urbanucci, L.; D'Ettorre, F. Synthesis and Optimal Operation of Smart Microgrids Serving a Cluster of Buildings on a Campus with Centralized and Distributed Hybrid Renewable Energy Units. *Energies* **2019**, *12*, 745. [CrossRef]
6. Péan, T.Q.; Salom, J.; Costa-Castelló, R. Review of control strategies for improving the energy flexibility provided by heat pump systems in buildings. *J. Process Control* **2019**, *74*, 35–49. [CrossRef]

7. D'Ettorre, F.; Conti, P.; Schito, E.; Testi, D. Model predictive control of a hybrid heat pump system and impact of the prediction horizon on cost-saving potential and optimal storage capacity. *Appl. Therm. Eng.* **2019**, *148*, 524–535. [CrossRef]
8. D'Ettorre, F.; de Rosa, M.; Conti, P.; Testi, D.; Finn, D. Mapping the energy flexibility potential of single buildings equipped with optimally-controlled heat pump, gas boilers and thermal storage. *Sustain. Cities Soc.* **2019**, *50*, 101689. [CrossRef]
9. Fischer, D.; Wirtz, T.; Zerbe, K.D.; Wille-Haussmann, B.; Madani, H. Test cases for hardware in the loop testing of air to water heat pump systems in a smart grid context. In Proceedings of the International Congress of Refrigeration (ICR), Yokohama, Japan, 16–22 August 2015; pp. 3666–3673.
10. Mehrfeld, P.; Nürenberg, M.; Knorr, M.; Schinke, L.; Beyer, M.; Grimm, M.; Lauster, M.; Müller, D.; Seifert, J.; Stergiaropoulos, K. Dynamic evaluations of heat pump and micro combined heat and power systems using the hardware-in-the-loop approach. *J. Build. Eng.* **2020**, *28*, 101032. [CrossRef]
11. Franco, A.; Fantozzi, F. Experimental analysis of a self consumption strategy for residential building: The integration of PV system and geothermal heat pump. *Renew. Energy* **2016**, *86*, 1075–1085. [CrossRef]
12. Beck, T.; Kondziella, H.; Huard, G.; Bruckner, T. Optimal operation, configuration and sizing of generation and storage technologies for residential heat pump systems in the spotlight of self-consumption of photovoltaic electricity. *Appl. Energy* **2017**, *188*, 604–619. [CrossRef]
13. El-Baz, W.; Mayerhofer, L.; Tzscheutschler, P.; Wagner, U. Hardware in the loop real-time simulation for heating systems: Model validation and dynamics analysis. *Energies* **2018**, *11*, 3159. [CrossRef]
14. Frison, L.; Kleinstück, M.; Engelmann, P. Model-predictive control for testing energy flexible heat pump operation within a Hardware-in-the-Loop setting. *J. Phys. Conf. Ser.* **2019**, *1343*, 012068. [CrossRef]
15. Schneider, G.F.; Oppermann, J.; Constantin, A.; Streblow, R.; Müller, D. Hardware-in-the-Loop-Simulation of a Building Energy and Control System to Investigate Circulating Pump Control Using Modelica. In Proceedings of the 11th International Modelica Conference, Versailles, France, 21–23 September 2015; Volume 118, pp. 225–233.
16. Hallerbach, S.; Xia, Y.; Eberle, U.; Koester, F. Simulation-Based Identification of Critical Scenarios for Cooperative and Automated Vehicles. *SAE Int. J. Connect. Autom. Veh.* **2018**, *1*, 1066. [CrossRef]
17. Tejeda de la Cruz, A.; Riviere, P.; Marchio, D.; Cauret, O.; Milu, A. Hardware in the loop test bench using Modelica: A platform to test and improve the control of heating systems. *Appl. Energy* **2017**, *188*, 107–120. [CrossRef]
18. Bianchi, M.; Shafai, E.; Federal, S.; Geering, H.P. Comparing New Control Concepts for Heat Pump Heating Systems on a Test Bench With the Capability of House and Earth Probe Emulation. In Proceedings of the 8th International Energy Agency Heat Pump Conference, Las Vegas, NV, USA, 30 May–2 June 2005; pp. 1–11.
19. Stutterecker, W.; Schoberer, T.; Steindl, G. Development of a Hardware-in-the-Loop Test Method for Heat Pumps and Chillers. In Proceedings of the REHVA Annual Conference "Advanced HVAC and Natural Gas Technologies", Riga, Latvia, 6–9 May 2015; Riga Technical University: Riga, Latvia, 2015; p. 117.
20. Pratt, A.; Ruth, M.; Krishnamurthy, D.; Sparn, B.; Lunacek, M.; Jones, W.; Mittal, S.; Wu, H.; Marks, J. Hardware-in-the-loop simulation of a distribution system with air conditioners under model predictive control. *IEEE Power Energy Soc. Gen. Meet.* **2018**, *2018*, 1–5.
21. Rhee, K.N.; Yeo, M.S.; Kim, K.W. Evaluation of the control performance of hydronic radiant heating systems based on the emulation using hardware-in-the-loop simulation. *Build. Environ.* **2011**, *46*, 2012–2022. [CrossRef]
22. EN 15316-4-2. *Heating Systems in Buildings—Method for Calculation of System Energy Requirements and System Efficiencies—Part 4-2: Space Heating Generation Systems, Heat Pump Systems*; European Committee for Standardization (CEN): Brussels, Belgium, 2017.
23. EN 14511-2. *Air Conditioners, Liquid Chilling Packages and Heat Pumps for Space Heating and Cooling and Process Chillers, with Electrically Driven compressors—Part 2: Test Conditions*; European Committee for Standardization (CEN): Brussels, Belgium, 2018.
24. EN 14825. *Air Conditioners, Liquid Chilling Packages and Heat Pumps, with Electrically Driven Compressors, for Space Heating and Cooling. Testing and Rating at Part Load Conditions and Calculation of Seasonal Performance*; European Committee for Standardization (CEN): Brussels, Belgium, 2018.

25. EN 52016-1. *Energy Performance of Buildings—Energy Needs for Heating and Cooling, Internal Temperatures and Sensible and Latent Heat Loads—Part 1: Calculation Procedures*; European Committee for Standardization (CEN): Brussels, Belgium, 2017.
26. Schito, E.; Conti, P.; Urbanucci, L.; Testi, D. Multi-objective optimization of HVAC control in museum environment for artwork preservation, visitors' thermal comfort and energy efficiency. *Build. Environ.* **2020**, *180*, 107018. [CrossRef]
27. EN 15316-1. *Energy Performance of Buildings—Method for Calculation of System Energy Requirements and System Efficiences—Part 1: General and Energy Performance Expression*; European Committee for Standardization (CEN): Brussels, Belgium, 2017.
28. EN 1397. *Heat Exchangers—Hydronic Room Fan Coil Units—Test Procedures for Establishing the Performance*; European Committee for Standardization (CEN): Brussels, Belgium, 2015.
29. Guoyuan, M.; Qinhu, C.; Yi, J. Experimental investigation of air-source heat pump for cold regions. *Int. J. Refrig.* **2003**, *26*, 12–18. [CrossRef]
30. Madonna, F.; Bazzocchi, F. Annual performances of reversible air-to-water heat pumps in small residential buildings. *Energy Build.* **2013**, *65*, 299–309. [CrossRef]
31. Testi, D.; Schito, E.; Tiberi, E.; Conti, P.; Grassi, W. Building Energy Simulation by an In-house Full Transient Model for Radiant Systems Coupled to a Modulating Heat Pump. *Energy Procedia* **2015**, *78*, 1135–1140. [CrossRef]
32. Dongellini, M.; Morini, G.L. On-off cycling losses of reversible air-to-water heat pump systems as a function of the unit power modulation capacity. *Energy Convers. Manag.* **2019**, *196*, 966–978. [CrossRef]
33. Tassou, S.A.; Votsis, P. Transient response and cycling losses of air-to-water heat pump systems. *Heat Recover. Syst. CHP* **1992**, *12*, 123–129. [CrossRef]
34. Bagarella, G.; Lazzarin, R.M.; Lamanna, B. Cycling losses in refrigeration equipment: An experimental evaluation. *Int. J. Refrig.* **2013**, *36*, 2111–2118. [CrossRef]
35. Piechurski, K.; Szulgowska-Zgrzywa, M.; Danielewicz, J. The impact of the work under partial load on the energy efficiency of an air-to-water heat pump. *E3S Web Conf.* **2017**, *17*, 00072. [CrossRef]

Article

Hybrid Fuel Cell—Supercritical CO_2 Brayton Cycle for CO_2 Sequestration-Ready Combined Heat and Power

Rhushikesh Ghotkar [1], Ellen B. Stechel [2], Ivan Ermanoski [3] and Ryan J. Milcarek [1,*]

1 School for Engineering of Matter, Transport and Energy, Arizona State University, Tempe, AZ 85287-6106, USA; rghotkar1@asu.edu
2 ASU LightWorks® and School of Molecular Sciences, Arizona State University, Tempe, AZ 85287-5402, USA; Ellen.Stechel@asu.edu
3 ASU LightWorks® and School of Sustainability, Arizona State University, Tempe, AZ 85287-5402, USA; Ivan.Ermanoski@asu.edu
* Correspondence: Ryan.Milcarek@asu.edu; Tel.: +1-480-965-2724

Received: 23 July 2020; Accepted: 22 September 2020; Published: 24 September 2020

Abstract: The low prices and its relatively low carbon intensity of natural gas have encouraged the coal replacement with natural gas power generation. Such a replacement reduces greenhouse gases and other emissions. To address the significant energy penalty of carbon dioxide (CO_2) sequestration in gas turbine systems, a novel high efficiency concept is proposed and analyzed, which integrates a flame-assisted fuel cell (FFC) with a supercritical CO_2 (sCO_2) Brayton cycle air separation. The air separation enables the exhaust from the system to be CO_2 sequestration-ready. The FFC provides the heat required for the sCO_2 cycle. Heat rejected from the sCO_2 cycle provides the heat required for adsorption-desorption pumping to isolate oxygen via air separation. The maximum electrical efficiency of the FFC sCO_2 turbine hybrid (FFCTH) without being CO_2 sequestration-ready is 60%, with the maximum penalty being 0.68% at a fuel-rich equivalence ratio (Φ) of 2.8, where Φ is proportional to fuel-air ratio. This electrical efficiency is higher than the standard sCO_2 cycle by 6.85%. The maximum power-to-heat ratio of the sequestration-ready FFCTH is 233 at a Φ = 2.8. Even after including the air separation penalty, the electrical efficiency is higher than in previous studies.

Keywords: supercritical CO_2; combined heat and power; flame-assisted fuel cells; carbon sequestration; solid oxide fuel cell

1. Introduction

Over the past few decades, natural gas (NG) production in the USA has increased by 40% [1]. NG has been termed a "bridge fuel" between the fossil carbon-intensive electric grid of today and the low fossil carbon grid of the future [2–4]. Along with less carbon dioxide (CO_2) emissions per kWh [5,6], switching from coal to NG also provides several health benefits. Natural gas power plants emit less sulfur dioxide (SO_2) [7], nitrogen oxides (NO_x) [2], and primary particulate matter [2] when compared to coal-fired power plants. Emissions of primary particulate matter ($PM_{2.5}$ and PM_{10}) have been linked to human mortality and morbidity [8–12]. Recent regulations have focused attention on reducing emissions and are drivers for a switch from coal to NG power plants [1,13].

Though NG plants have much lower emissions than coal, they still produce substantial amounts of CO_2—one of the most prominent greenhouse gas other than water vapor in the atmosphere [14]. Therefore, to minimize CO_2 emissions, while still maintaining dispatchability, it is important to address two aspects of power generation: (1) increasing power plant efficiency, and (2) sequestering CO_2.

An electrical efficiency frontrunner power cycle is the supercritical CO_2 (sCO_2) Brayton cycle. Further improving the efficiency will require hybrid approaches, including topping and/or bottoming cycles; one example is the use of solid oxide fuel cells (SOFCs) as a topping cycle.

Dual chambered solid oxide fuel cells (DC-SOFC) are highly efficient, but suffer from limited fuel flexibility due to carbon coking with hydrocarbon fuels [15]. Fuel reformers or catalysts address coking but increase complexity and cost. DC-SOFC also have low thermal endurance leading to sealant failure and performance reduction under cycling load [16,17], so they are not a good candidate for long-term practical applications.

Flame-assisted fuel cells (FFC), and the related direct flame fuel cells (DFFCs) [18–22], were developed to overcome the limitations of DC-SOFC [23–29]. In the FFC setup, fuel-rich combustion (i.e., partial oxidation) generates syngas (i.e., H_2 and CO), which generates power in the SOFC [28]. Lean combustion of the remaining fuel maximizes heat recovery [28,30]. Integration of FFC with NG fueled combustion subsystems can lead to improvements in thermal cycling and an increase in the overall system efficiency while reducing the complexity and cost compared to a DC-SOFC system [25,29,31]. Recently, a paper investigating the integration of FFC with an air Brayton cycle has shown a significant increase in net electrical efficiency due to the integration [32]. Along the same lines, the FFC with a sCO_2 turbine, proposed in this paper, aims to achieve electrical efficiency gains at a lower cost compared to integration with a traditional SOFC system.

Further decrease in CO_2 emissions can be achieved by CO_2 sequestration, which has been investigated recently [33–35]. Integrating CO_2 sequestration with common power cycles leads to efficiency and cost penalties [36,37], which must be minimized. The approach we examine here relies on low temperature (323–473 K depending on the material) thermally-driven adsorption/desorption cycle for air separation [38,39] to provide pure oxygen to the power cycle. Using pure oxygen produces an exhaust stream containing only CO_2 and water vapor, which after condensing the water vapor, is sequestration-ready. In the activation step of the sorption cycle, a high surface area solid sorbent with adsorbed oxygen is heated, leading to O_2 desorption. In the pumping step, oxygen from air exothermically chemisorbs on the surface (either molecular or dissociative) [38]. Due to the low temperature requirement, the heat required for the activation step can make use of the heat rejected by the sCO_2 cycle. With CO_2 removal from the exhaust, the FFC-sCO_2 turbine hybrid (FFCTH) can be a zero carbon emissions power and heat generation system.

Here, we analyze the combined approach to emissions-free, CO_2 sequestration-ready, NG power generation, via the integration of a FFC and sorption air separation, with a sCO_2 Brayton cycle. The FCC and sCO_2 turbine generates power; heat recovery makes this FFCTH a combined heat and power (CHP) system. This analysis investigates the benefits of the FFCTH integration showing a plausible path to zero emissions, high efficiency, combined heat, and power.

2. Theory

This section establishes the theory of the overall concept and the individual components in the FFCTH. We first show a detailed analysis of the contribution from each component in the overall system. Following this component analysis, we provide a description of the experimental setup for testing the FFC along with the materials and methods used to conduct the experiment. Lastly, we use the results from the experiments in the theoretical analysis to evaluate the concept further.

2.1. Theoretical Basis: sCO_2 Cycle, FFCs, Air Separation, and System Level

2.1.1. Standard sCO_2 Brayton Cycle

This section reviews the theory of a standard sCO_2 Brayton cycle with recuperation and recompression, to provide a baseline for performance of a standalone sCO_2 cycle. Figure 1 shows a schematic of a standard sCO_2 cycle with recuperation and recompression [40]. At state 1, 53% of CO_2 at a pressure of 7.5 MPa and a temperature of 300 K enters the system where it is compressed to a higher pressure and reaches state 2. Meanwhile, the secondary compressor compresses the rest of the CO_2, which starts at the same pressure, but higher temperature and reaches state 2a.

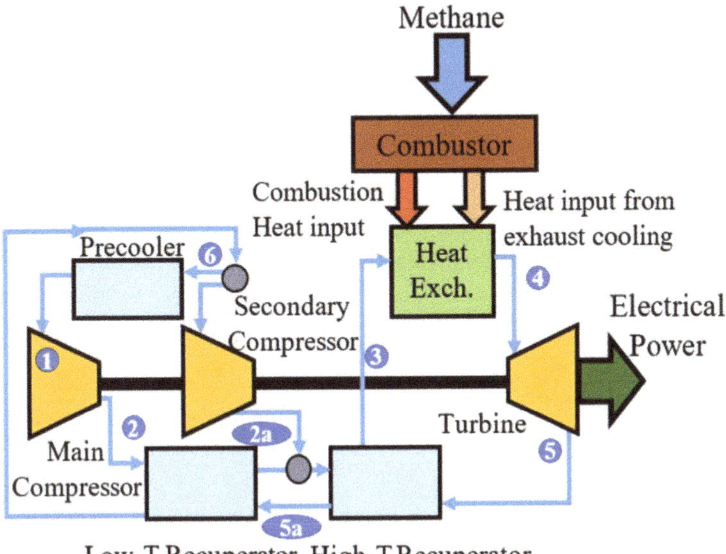

Figure 1. Schematic of a standard sCO_2 Brayton cycle with recuperation and recompression showing the various state points in the system.

The initial CO_2 from state 2 is preheated in the low temperature recuperator and reaches state 2a where the two CO_2 streams combine. The complete stream is then further preheated in the high temperature recuperator and reaches state 3. Further heat addition from external sources like combustion of a fuel or a concentrating solar thermal heat transfer fluid via a heat exchanger brings the CO_2 to state 4. For the purposes of the analysis here, we use combustion of methane to generate the heat in order to provide the most direct comparison with the FFCTH system. A turbine extracts mechanical energy from the sCO_2 stream between state 4 to state 5. The turbine exit stream preheats the incoming CO_2 streams in the high temperature recuperator, reaching state 5a, and in the low temperature recuperator, reaching state 6. Heat rejection to the environment or for external processing heating occurs in the pre-cooler, to return to state 1. Cooling the exhaust from the combustor and supplying thermal energy to the sCO_2 turbine cycle facilitates maximizing the electrical efficiency.

The electrical efficiency of the standard sCO_2 cycle (η_{SSGT}) given in Figure 1 is represented in Equation (1), where $\dot{m}_{CO_2}$ is the total CO_2 mass flow rate in the cycle. The specific enthalpies of states 1, 2, 2a, 4, 5, and 6, are h_1, h_2, h_{2a}, h_4, h_5, and h_6, respectively. The coefficients 0.53 and 0.47 are the mass fraction of CO_2 entering the main compressor and secondary compressor, respectively. The mass flow rate of methane required to provide the necessary thermal energy is $\dot{m}_f$, and HHV_f is the higher heating value of methane. P_{CC} is the total electrical power required for CO_2 compression during sequestration. Previous literature has shown that the specific power required for compression and refrigeration of CO_2 for sequestration at 15 Bar is 107 kW/kg CO_2 [41], which is used in this work. This power is kept zero when sequestration is not assessed.

$$\eta_{SSGT} = \frac{\dot{m}_{CO_2}[(h_4 - h_5) - 0.47(h_{2a} - h_6) - 0.53(h_2 - h_1)] - P_{CC}}{\dot{m}_f HHV_f} \quad (1)$$

2.1.2. Flame-Assisted Fuel Cells

This section provides a theoretical model of the FFC, in which performance depends on the fuel-rich equivalence ratio (Φ), defined in Equation (2), as the independent variable. A more detailed FFC model is given in [42]. Here, n_{CH_4} and n_{o_2} are molar flow rates of methane and oxygen, respectively. The superscript 'S' denotes rates required for stoichiometric reaction. Thus, combustion is fuel-rich for $\Phi > 1$, fuel-lean for $\Phi < 1$, and stoichiometric for $\Phi = 1$.

$$\Phi = \frac{\frac{n_{CH_4}}{n_{O_2}}}{\frac{n^s{}_{CH_4}}{n^s{}_{O_2}}} \tag{2}$$

Figure 2 shows a schematic of a FFC with various reaction zones. The SOFC in the FFC configuration consists of a porous anode and cathode separated by a dense electrolyte layer. Partial oxidation of the fuel (i.e., methane in this study) and oxygen mixture sent to the fuel-rich pre-burner results in the generation of syngas (H_2 + CO). The syngas then diffuses into the anode where it reacts with the oxygen ions diffusing from the cathode side through the electrolyte to form CO_2 and water. After the fuel cell, remaining syngas combusts with oxygen, in the fuel-lean (excess oxygen reactant) after-burner and generates heat. The after-burner exhaust exits the fuel cell subsystem. The overall FFC subsystem generates heat during fuel-rich combustion, fuel cell electrochemical oxidation, and fuel-lean combustion.

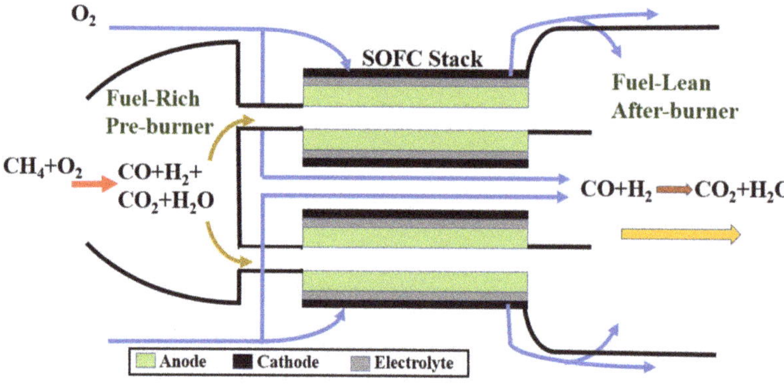

Figure 2. Schematic of a flame-assisted fuel cell (FFC).

The first reaction is the fuel-rich combustion of methane in oxygen. Equation (3) shows this reaction. Chemical equilibrium from a Gibbs minimization constrained by conservation of the elements determines the stoichiometry of the products for fuel-rich combustion, i.e., *a*, *b*, *c*, and *d*, for CO, H_2, CO_2, and H_2O, respectively.

$$\Phi CH_4 + 2(O_2) \rightarrow a CO + b\, H_2 + c\, CO_2 + d\, H_2O \tag{3}$$

The enthalpy released by the fuel-rich combustion reaction (ΔH_{RC}) can be calculated as shown in Equation (4), where $\dot{m}_i$ is the mass flow rate for species i and h_i is the species specific enthalpy (i.e., per unit mass).

$$\Delta H_{RC} = \dot{m}_{CO} h_{CO} + \dot{m}_{CO_2} h_{CO_2} + \dot{m}_{H_2O} h_{H_2O} + \dot{m}_{H_2} h_{H_2} - \dot{m}_{CH_4} h_{CH_4} \tag{4}$$

Various losses give rise to three key efficiency definitions in the FFC subsystem. Those include the fuel utilization efficiency (η_{fu}), the fuel cell conversion efficiency (η_{fc}), and the overall efficiency (η_{ov}). Equations (5)–(7) define each efficiency, respectively.

$$\eta_{fu} = \frac{\text{Syngas electrochemically oxidized in FFC}}{\text{Total Syngas available in exhaust}} \tag{5}$$

$$\eta_{fc} = \frac{\text{Electrical power generated by FFC}}{\text{Total Chemical energy of the available syngas}} \tag{6}$$

$$\eta_{ov} = \frac{\text{Electrical power generated by fuel cell}}{\text{Chemical energy from the hydrocarbon fuel}} \tag{7}$$

FFC electrochemically oxidized both H_2 and CO to generate electric power. Equation (8) shows the effective fuel cell reaction), in which the coefficients a_1, b_1, and γ_1 depend upon the a and b stoichiometric coefficients from Equation (3) and the fuel utilization efficiency of the FFC (Equation (5)).

$$a_1\ CO + b_1\ H_2 + \gamma_1 O_2\ \rightarrow a_1\ CO_2 + b_1\ H_2O \tag{8}$$

Writing the reaction Equation (8) per mole of syngas leads to Equation (9).

$$\frac{a_1}{a_1+b_1}\ CO + \frac{b_1}{a_1+b_1}\ H_2 + \frac{\gamma_1}{a_1+b_1}\ O_2\ \rightarrow \frac{a_1}{a_1+b_1}\ CO_2 + \frac{b_1}{a_1+b_1}\ H_2O \tag{9}$$

The mole specific standard Gibbs' free energy released by reaction Equation (8) (Δg°_{FC}) and the temperature is used to calculate the reversible cell potential (V_{rev}) of the fuel cell as shown in Equation (10). In Equation (10), R is the universal gas constant, T is absolute temperature (in Kelvin), and K is the equilibrium constant of reaction in Equation (8). The mole specific enthalpy of reaction (8) (Δh_{FC}) is used to calculate the thermo-neutral potential (V_{th}) as shown in Equation (11) where n is the number of electrons released per mole of fuel in the fuel cell reaction (two electrons per mole of syngas) and F is Faraday's constant.

$$V_{rev} = \frac{-\Delta g^{o}{}_{FC}}{nF} - \frac{RT}{nF}\ \ln(K) \tag{10}$$

$$V_{th} = \frac{-\Delta h_{FC}}{nF} \tag{11}$$

The equilibrium constant is calculated as shown in Equation (12) where P_i/P is the ratio of partial pressure of species i and y_i/y_{syn} is the ratio of coefficients of species i in reaction Equation (8). The ratio of partial pressures in the subsystem is equal to the mole fraction of the components, assuming ideal gas behavior.

$$K = \prod_{i=1}^{N}\left(\frac{P_i}{P}\right)^{\frac{y_i}{y_{syn}}} \tag{12}$$

For the fuel cell reaction, Equation (13) shows how to determine the equilibrium constant, where X_i is the mole fraction of species 'i'.

$$K = X_{H_2}^{\left(\frac{-b}{a+b}\right)} X_{O_2}^{\left(\frac{-\gamma}{a+b}\right)} X_{CO}^{\left(\frac{-a}{a+b}\right)} \tag{13}$$

Equation (14) shows the power generated by the fuel cell (P_{fc}), where η_{fu} is the fuel utilization efficiency, η_{fc} is the fuel cell efficiency, $\dot{n}_k$ is the molar flow rate of species 'k' (CO or H_2) in the fuel-rich exhaust, $\Delta g_{CO,CO_2}$ and $\Delta g_{H_2,H_2O}$ are the mole specific Gibbs' free energies released from the oxidation of CO and H_2, respectively.

$$P_{fc} = -\eta_{fu}\eta_{fc}\left(\dot{n}_{CO}\Delta g_{CO,CO_2} + \dot{n}_{H_2}\Delta g_{H_2,H_2O}\right) \tag{14}$$

Equation (15) shows the total heat released by the combined fuel cell reactions; $\Delta h_{CO,CO_2}$ and $\Delta h_{H_2,H_2O}$ are the mole specific enthalpy changes from the oxidation of CO and H_2, respectively.

$$H_{fc} = -\eta_{fu}\cdot(1-\eta_{fc})\cdot(\dot{n}_{CO}\Delta h_{CO,CO_2} + \dot{n}_{H_2}\Delta h_{H_2,H_2O}) \tag{15}$$

After the fuel cell, the remaining fuel passes into the fuel-lean combustion chamber to generate more thermal energy for transfer to the sCO_2 stream via an indirect heat exchanger. Equation (16) shows the fuel-lean combustion reaction. In Equation (16), the stoichiometric coefficients a_2, b_2, and γ_2 depend on stoichiometry a and b from Equation (3) and on the η_{fu}. Thus, if the fuel-rich combustion produces b moles of H_2 and a moles of CO and b_1 and a_1 moles of H_2 and CO, respectively, react in the fuel cell reaction, then a_2 equals the difference between a and a_1 and b_2 equals the difference between b and b_1. Fixing the Φ of the fuel-lean combustion determines the value of γ_2. The assumption and expectation is that CO_2 and H_2O generated in the fuel-rich combustion, FFC, and the fuel-lean combustion reactions remain unreacted downstream of when they are produced.

$$a_2\ CO + b_2\ H_2 + \gamma_2 O_2 \rightarrow a_2\ CO_2 + b_2\ H_2O \tag{16}$$

2.1.3. FFC sCO_2 Turbine Hybrid with and without Carbon Sequestration

Figure 3 shows a schematic of the proposed FFCTH system, where the sCO_2 turbine receives heat input from the FFC via an indirect heat exchanger. The optional air separation unit provides oxygen to the FFC and after-burner, with the resulting exhaust requiring only water removal to be sequestration-ready.

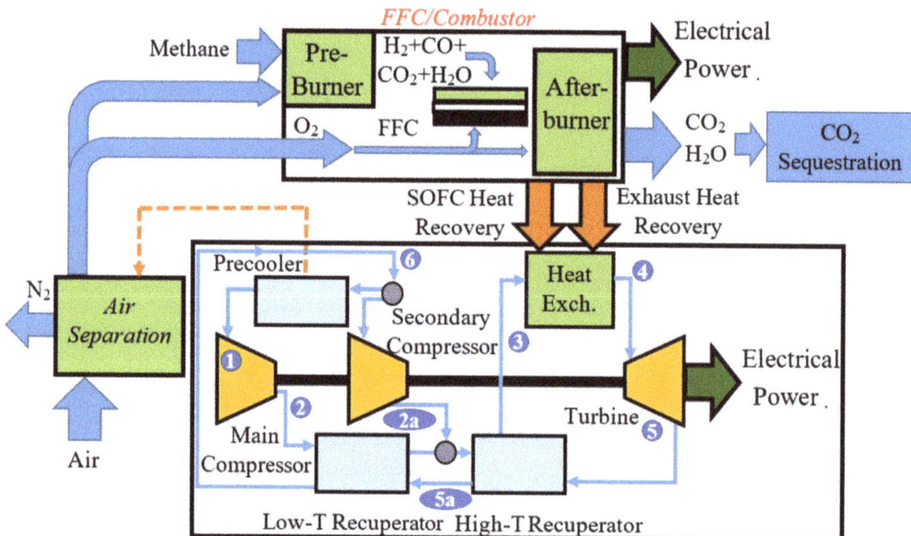

Figure 3. Schematic of the proposed FFC integrated sCO_2 turbine hybrid.

Use of FFC, instead of a conventional DC-SOFC, significantly reduces system complexity and size. The FFCTH has several tunable parameters, such as Φ, the FFC operating voltage/current, the air separation unit, and sCO_2 cycle variables. Examples of the latter are the pressure ratio and the turbine inlet temperature. All these tunable variables enable optimizing for a high power-to-heat ratio (P/H), high electrical efficiencies, and/or high thermal efficiencies. One key advantage of the system proposed here is that the heat rejected from the pre-cooler can drive the sorption/desorption based air separation unit, thus making this system self-sustained and powered by one fuel source and one injection point. The efficiency of the sorption/desorption based air separation unit is assumed to be in the acceptable range. This assumption applies due to the various tunable state points in the cycle and parameters in the system depending upon the materials used for air separation. The state points of the sCO_2 cycle remain the same as shown in Figure 1 with two additional components: (1) heat recovery from the FFC subsystem and (2) air separation from a sorption unit.

To identify the advantages and drawbacks of redesigning so that the system can be CO_2 sequestration-ready, we theoretically analyzed this system with and without air separation. Without separation, the FFC subsystem uses air, whereas with separation it produces and utilizes pure oxygen. We compare the system efficiencies, the fuel requirement, and the P/H for cases with and without the FFC integrated and with and without sorption-based air separation. Equation (17) shows the electrical efficiency of the FFC integrated sCO_2 turbine FFCTH (η_{FFCGT}), where P_{FFC} and P_{CO_2GT} is the power generated by FFC and by the sCO_2 turbine cycle, respectively. The heat required for the air separation unit is taken from the heat rejected in the sCO_2 cycle pre-cooler (states 6 to 1). The electrical efficiency of the FFCTH is given by Equation (17).

$$\eta_{FFCGT} = \frac{P_{FFC} + P_{CO_2GT} - P_{CC}}{\dot{m}_f HHV_f} \tag{17}$$

3. Experimental

An experimental assessment and analysis of the FFC performance operating in methane and oxygen fuel-rich combustion exhaust provided data to calibrate and enable simulation of the FFCTH system. Figure 4 shows the schematic of the experimental setup. Use of mass flow controllers regulated the flow of the gases. Connecting the fuel lines for CO and H_2 to flame arrestors mitigates and avoids flashback risk.

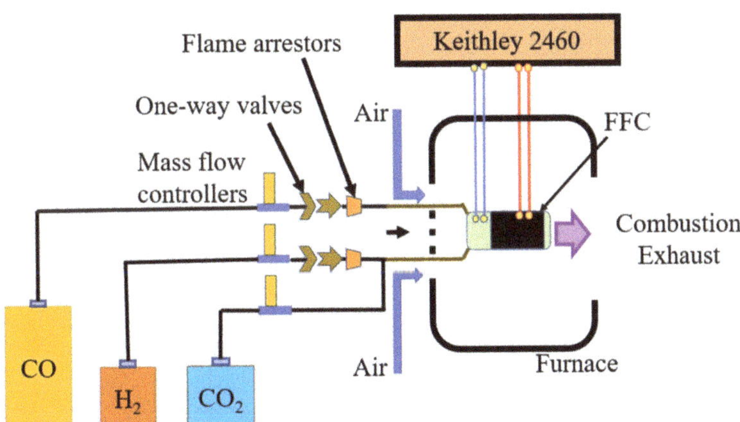

Figure 4. Schematic of the experimental setup with gases mixture used for model exhaust.

The flow rates of the gases used in the experiment (CO, H_2, CO_2) were calculated using chemical equilibrium, from NASA Chemical Equilibrium and Applications (CEA) [43], for the combustion exhaust composition at a constant inlet fuel flow rate of 0.066 mg·s^{-1}. Methane was used as an approximation for natural gas since natural gas is up to 90% methane by molar content [44] and thus, using methane instead of natural gas would not lead to significant departure from the natural gas case. Sulfur impurities will need to be removed from the natural gas before its use in the fuel cell due to poisoning effect on the anode. Other hydrocarbon impurities are not expected to damage the fuel cell since they are reformed in the fuel-rich pre-burner. Tables 1 and 2 show the NASA CEA exhaust composition for combustion of methane in air and methane in oxygen, respectively. The term used for this exhaust composition, when obtained from chemical equilibrium, is 'model fuel-rich exhaust'. The case of methane in air serves as a base case for comparison; however, the experiment used only the model fuel-rich exhaust for methane in oxygen as Table 2 shows. Since it is difficult to work with steam, and since steam should be non-reactive electrochemically, increasing the CO_2 molar flow serves to eliminate while compensating steam from the experiment. Thus, CO_2 molar flow rate in the experiment was the same total (steam plus CO_2) as in the proposed concept.

Table 1. NASA CEA exhaust composition of methane and air for different Φ.

Φ	CO (mL·min^{-1})	H_2 (mL·min^{-1})	CO_2 (mL·min^{-1})	H_2O (mL·min^{-1})	N_2 (mL·min^{-1})
1.20	0.05	0.03	0.05	0.18	0.66
1.40	0.08	0.06	0.04	0.17	0.63
1.60	0.10	0.09	0.03	0.16	0.60
1.80	0.12	0.13	0.02	0.15	0.57
2.00	0.13	0.17	0.02	0.13	0.55
2.20	0.14	0.20	0.02	0.11	0.53
2.40	0.15	0.23	0.01	0.09	0.51
2.60	0.16	0.26	0.01	0.08	0.49
2.80	0.17	0.29	0.01	0.06	0.47

Table 2. NASA CEA exhaust composition of methane and oxygen for different Φ.

Φ	CO (mL·min^{-1})	H_2 (mL·min^{-1})	CO_2 (mL·min^{-1})	Total (mL·min^{-1})
1.20	0.2672	0.1560	0.0848	0.4715
1.40	0.2855	0.2000	0.0695	0.4333
1.60	0.2981	0.2503	0.0555	0.3899
1.80	0.3061	0.3053	0.0435	0.3418
2.00	0.3110	0.3614	0.0341	0.2921
2.20	0.3141	0.4142	0.0272	0.2444
2.40	0.3163	0.4612	0.0222	0.2010
2.60	0.3182	0.5020	0.0185	0.1625
2.80	0.3201	0.5369	0.0158	0.1289

Equation (18) shows the formula to calculate the volumetric flow rate of species '*i*' as a function of Φ. In Equation (18), X_i and X_{CH4} are the mole fractions of species '*i*' and methane, respectively, obtained from Table 2 at the respective Φ. V_{mol} is the molar volume of an ideal gas at 298 K, 1 bar pressure. Table 3 shows the flow rates subsequently obtained. The gases were all mixed and sent to the FFC subsystem inside a furnace at 1073 K. Previous experiments have shown that the ionic and electronic conductivity of the electrodes and the ionic conductivity of the electrolyte are high at a temperature of 1073 K and above [16]. Furthermore, carbon deposition due to the carbon monoxide disproportionation reaction ($2CO \rightarrow C + CO_2$) is less favorable at operating temperatures of 1073 K and higher [16]. Thus, a temperature of 1073 K was chosen for the experiments. Air in the furnace was supplied to the fuel cell cathode.

$$V_i = \frac{\dot{n}_f}{X_{CH_4}} \times X_i \times V_{mol} \tag{18}$$

Table 3. Flow rates of gases used in the experiment at different Φ.

Φ	CO (mL·min^{-1})	H_2 (mL·min^{-1})	CO_2 (mL·min^{-1})	Total (mL·min^{-1})
1.20	3.33	1.94	7.18	12.45
1.40	3.81	2.68	6.87	13.36
1.60	4.22	3.55	6.40	14.17
1.80	4.55	4.55	5.78	14.88
2.00	4.82	5.59	5.06	15.47
2.20	5.00	6.61	4.33	15.94
2.40	5.15	7.51	3.63	16.29
2.60	5.25	8.28	2.96	16.49
2.80	5.32	8.93	2.38	16.63

Fuel Cell Fabrication

Fabrication methods for the FFC anode (NiO + YSZ, $(Y_2O_3)_{0.08}(ZrO_2)_{0.92}$) and the electrolyte (YSZ, ~22 μm thick) used in this study were reported previously in the literature [24]. Pre-firing of the anode occurred at 1373 K. The electrolyte was dip coated on the anode and sintered at 1673 K for four hours. A buffer layer of $Sm_{0.20}Ce_{0.80}O_{2-x}$ (SDC) was deposited onto the electrolyte using spray deposition [45]. An SDC+LSCF $(La_{0.6}Sr_{0.4})_{0.95}Co_{0.2}0Fe_{0.8}O_{3-x})$ cathode was deposited onto the buffer layer using dip coating, later dried and sintered at 1373 K for two hours. The final internal diameter of the tubular FFC is 2.2 mm and the outer diameter is 3.3 mm. The current collectors on the cathode and anode use silver wire and gold paste. The total active area of the cell is 4.32 cm^2. A source meter (Keithley 2460) is connected to the anode and cathode current collector. We used the current-voltage method with a four-probe technique to obtain the polarization curve and the power density.

4. Results and Discussion

4.1. Fuel Cell Performance

Figure 5 shows the performance of the FFC operating at 1073 K with a simulated methane/oxygen fuel-rich combustion exhaust composition between the Φ of 1.2 to 2.8 and flow rates shown in Table 3. A maximum Φ of 2.8 was chosen as carbon formation becomes thermodynamically favorable at higher Φ and peak hydrogen concentration occurs near this Φ. Figure 5 shows that significant power densities were achieved at all Φ.

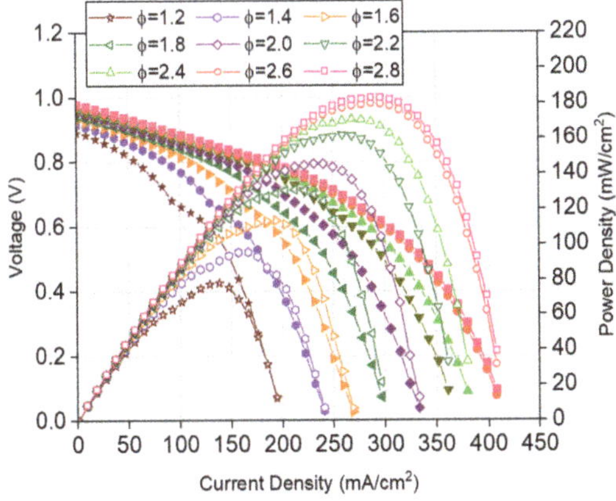

Figure 5. Modeling results for the FFC operating voltages and power density at 1073 K using a model fuel-rich exhaust composition between 1.2 and 2.8.

The significant change in FFC power density shown in Figure 5 is due to a change in polarization at different Φ. At lower current densities (<100 mA/cm^2), the slope of the polarization curves remains similar at all Φ indicating dominant ohmic losses. At higher current densities, the polarization losses increase as Φ decreases. This happens because as the Φ decreases, the concentration of syngas in the combustion exhaust decreases, as Table 2 shows, leading to higher concentration losses. There are no significant activation losses at any Φ. As a result of changes in polarization, the power density at each current density increases as the Φ increases. As the Φ increases, more syngas is available in the combustion exhaust leading to larger Gibbs' free energy released, which in turn leads to less concentration losses and more total power being generated in the FFC. At an operating voltage of 0.6 V, the maximum power density of 183 mW/cm^2 occurs at a Φ of 2.8. This power density result (<200 mW/cm^2) may not seem impressive on its own; however, consideration of the fuel-utilization efficiencies achieved during the experiment does give it more significance.

The fuel utilization efficiencies reached during the experiment are significantly higher than those found in literature for FFCs [16,24,42,46]. For example, a fuel utilization of 75% at Φ of 1.2 and 63% at Φ of 2.8 is obtained at an operating voltage of 0.6 V. Low flow rates and the controlled experiment in the furnace are the primary reasons for the high fuel utilization observed. Experiments were also conducted for a FFC operating in methane/air fuel-rich combustion exhaust and similar operating characteristics, fuel utilization, and fuel cell efficiency were obtained. The model, presented below, uses experimental parameters of fuel utilization and fuel cell efficiency for the FFC operation in methane/oxygen and methane/air fuel-rich combustion exhaust.

It is important to consider the validity of the FFC model described earlier for prediction of experimental results. From the results in Figure 5, the FFC open circuit voltage shows a clear increasing trend with Φ due to decreased Nernstian loss, as predicted by Equation (10). Figure 6 shows that the analytically calculated reversible cell potential and the open circuit voltage increase with increase in Φ due to increase in the syngas concentration in the fuel-rich combustion exhaust. This comparison indicates that the FFC model is in approximate agreement (<5% variation at all Φ) with the experimental results validating the theoretical model described in the paper.

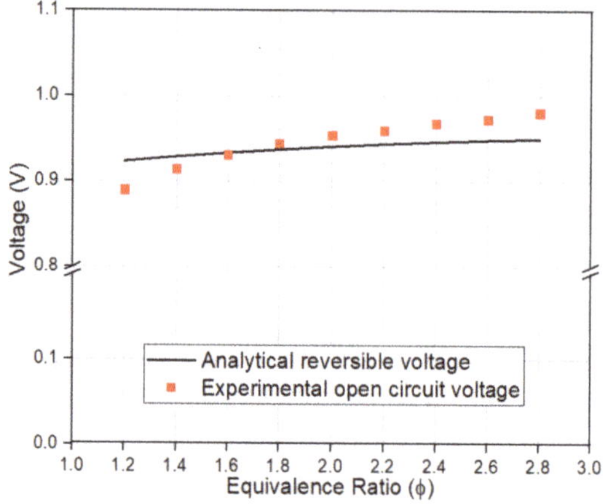

Figure 6. Comparison of analytically calculated reversible voltage and experimentally obtained open circuit voltage.

4.2. sCO_2 Turbine Performance with and without an Integrated FFC Subsystem

The computations for the FFCTH assumed that the FFC would achieve similar performance when scaled to design power as it did during the experiment. All thermochemical properties of gases were taken from NIST (Gaithersburg, MD, USA) software MINI-REFPROP [47].

To isolate the effects of integrating the FFC system with the standard sCO_2 cycle, the total power generated by the turbine system alone was held constant at 6 MW. To generate this power, Table 4 lists the temperature and pressure of various state points as defined in Figure 1.

Table 4. Temperatures and pressures of the state points in the sCO_2 cycle.

State Point	Temperature (K)	Pressure (MPa)
1	300	7.5
2	318	20
2a	414	20
3	733	20
4	913	20
5	780	7.5
5a	452	7.5
6	331	7.5

The state points in Table 4 were taken from a previous sCO_2 power cycle [40]. The mass flow rate of CO_2 required to generate 6 MW of power is 48.35 kg/s. The total amount of heat rejected from the sCO_2 turbine system is 4.81 MW. The compression ratio in the system is 2.66. With these parameters, we evaluated the performance of an integrated FFC system, which is able to provide the heat required by sCO_2 cycle. Air or oxygen is the oxidant for the FFC system. The oxygen case enables sequestration-ready exhaust from the FFC system. The heat rejected from the pre-cooler (states 6 to 1 in Figure 3) provides heat to separate oxygen from air using a thermally driven adsorption/desorption cycle [38]. If the heat rejected from the sCO_2 cycle is not enough to separate enough oxygen for the FFC system, we adjusted the FFC operating voltage to enable more heat rejection. Also note that the efficiency of the air-separation is assumed to be acceptable in the design. The flexibility of the cycle state points to give us this unique opportunity of analyzing this cycle and the assumption is thus justified. The state points will need to be modified depending upon the specific material used for adsorption–desorption pumping of oxygen for air separation [38,39]. The fixed parameters used in the analysis is shown in Table 5.

Table 5. Summary of the fixed parameters used in the analysis.

Property	Value
Power generated by standard supercritical CO_2 turbine cycle	6 MW
Efficiency of the standard supercritical CO_2 turbine cycle (based on states 1–6)	53.14%
Compression ratio	2.66

The fuel flowrate requirement of the FFCTH and its variation with Φ was assessed. The initial conditions for air (or oxygen) entering the fuel-rich combustion chamber of the FFC is 1 bar and 298 K. The Φ of the fuel-rich preburner varies, while Φ of the fuel-lean after-burner remains constant at 0.8. The assumed efficiency for the heat exchangers in the system is 90%. η_{fu} and η_{fc} (fixed at 0.7 and 0.5, respectively) are the average experimentally measured results over all Φ. Figure 7 shows the flow rate of methane required to generate the heat (10.8 MW) from the FFC with and without being CO_2 sequestration-ready. Figure 7 shows the base-case fuel-flow required to meet the heat requirements with a standard sCO_2 turbine cycle (Figure 1).

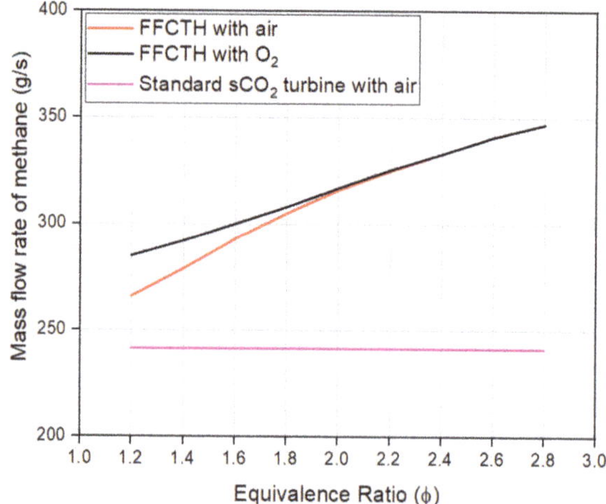

Figure 7. Flow rates of methane required to produce the required amount of heat for sCO_2 cycle with air and oxygen oxidizers for different Φ.

As shown in Figure 7, at all Φ the fuel flow required for the standard sCO_2 cycle is lower than the fuel flow for the FFCTH with and without being CO_2 sequestration-ready. In the standard sCO_2 cycle case, the fuels chemical energy is converted to heat and then to the sCO_2 cycle via the heat exchanger. Thus, the standard cycle requires less fuel, but also produces less power than in the case of the integrated FFC. The standard sCO_2 cycle operates at a fixed Φ of 0.8 as all of the heat released in the combustor transfers to the working fluid. This fuel flow result shows that more methane is required for the sequestration-ready FFCTH power generation compared to the sCO_2 cycle because the electrical power generated by the sCO_2 is fixed at 6 MW. However, the electrical efficiency of the FFCTH is higher overall, which will be shown below.

When integrating the FFC with the sCO_2 cycle, the amount of methane required to provide the necessary heat increases with Φ. This increase happens because as Φ increases a larger portion of the incoming fuel energy converts to electric power in the FFC due to the constant fuel utilization efficiency. To make up for the larger power generation, more fuel is necessary to meet the heat requirement. Thus, even though the FFCTH requires more methane flow to operate, it is important to consider the electrical efficiency (described in later sections) of the setup in order to establish the significance of these results. It also shows why the comparison of methane flow rate of FFCTH with standard sCO_2 turbine cycle alone can be misleading.

Figure 7 also shows that the fuel needed to meet the heat requirement is slightly higher with oxygen (CO_2 sequestration-ready case) than with air at lower Φ (7% higher at Φ = 1.2). At higher Φ (>2), the fuel flow rates converge to the same value. To understand the reason for this trend, it is important to consider the syngas concentration of the fuel-rich combustion exhaust with air compared to with oxygen. For the fuel-rich combustion exhaust concentrations for methane with air and oxygen, we refer back to Tables 1 and 2. To maintain consistency for the purpose of the comparison, the nitrogen is removed, and the rest of the concentrations are rescaled, so they sum to 1. Table 6 shows these scaled concentrations.

Table 6. NASA CEA fuel-rich exhaust compositions for methane in air scaled to 1 without nitrogen.

Φ	CO	H_2	CO_2	H_2O
1.20	0.17	0.09	0.17	0.57
1.40	0.23	0.16	0.11	0.50
1.60	0.26	0.24	0.08	0.42
1.80	0.28	0.31	0.06	0.35
2.00	0.29	0.38	0.04	0.29
2.20	0.30	0.43	0.03	0.24
2.40	0.31	0.47	0.03	0.19
2.60	0.31	0.51	0.02	0.16
2.80	0.31	0.54	0.02	0.12

Figure 8 shows the trend in the variation of scaled syngas concentrations in the fuel-rich exhaust for methane combustion in air and oxygen. It is evident that the trend in Figure 7 matches the trend in the syngas composition of the fuel-rich combustion exhaust. The syngas oxidizes electrochemically and produces power in the FFC at an assumed constant fuel utilization. Hence, when the flow rates are equal, the higher syngas concentration for the methane/oxygen case will result in more electricity generation due to the constant fuel utilization in the FFC. Higher electricity generation for the methane/oxygen case will require a high flow rate (shown in Figure 7) in order to meet the thermal energy requirement of the sCO_2 bottoming cycle.

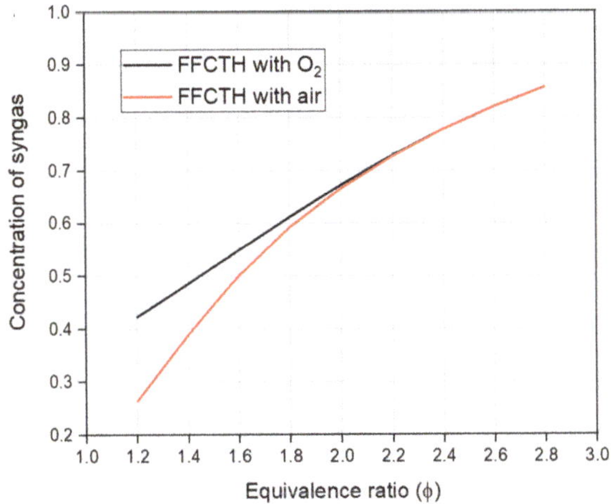

Figure 8. Syngas concentration in the fuel-rich combustion exhaust for methane combustion in air (without nitrogen and scaled to 1) compared to methane combustion in oxygen against increasing Φ.

Figure 9 shows the power generated by the sCO_2 cycle with oxygen and with air for the FFCTH. As a base case, Figure 9 shows the power generated by the standard sCO_2 turbine cycle. We note that the power generated in the oxygen case includes the electrical power penalty required for sequestration (i.e., compression) of the exhaust CO_2, as well as the air separation.

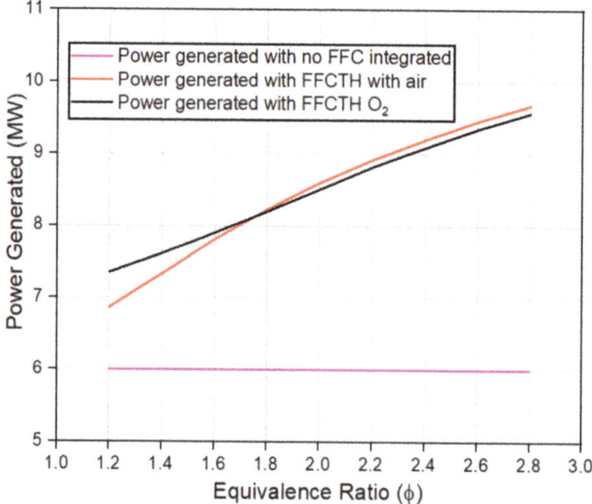

Figure 9. Comparison of the power generated by the sCO_2 turbine with and without FFC integrated and with and without being CO_2 sequestration-ready.

As shown in Figure 9, the power generated by the FFCTH increases with increasing Φ. This increase occurs because increasing the Φ increases the concentration of syngas in the fuel-rich combustion exhaust as shown in Table 1. Electrochemical oxidation of the syngas generates power; hence, more syngas leads to more electrical power generated as long as the fuel utilization remains high. Figure 5 shows the same trend for the experimental results, which confirms the trend and explains the model results.

The total power generated by the FFCTH system with and without being CO_2 sequestration-ready follows a slightly different trend than the one in Figure 7. This difference is primarily due to the compression penalty for making the exhaust CO_2 sequestration-ready. The power required for sequestration is small (0.1 MW) in comparison to the total power generated (~9.6 MW) at $\Phi = 2.8$. The total power generated by the FFCTH with being CO_2 sequestration-ready is 7% higher at $\Phi = 1.2$ and 1.2% lower at $\Phi = 2.8$ compared with no sequestration intent. It can also be seen that the total power generated by the FFCTH is significantly higher (~60% higher at $\Phi = 2.8$) compared to the standard sCO_2 turbine due to a large portion of the total power being generated by the FFC and the higher methane flow rate. Thus, analytical results show that the power generated by various cycles considered follows a slightly different trend than the trend for methane flow rate described earlier.

Figure 10 shows the variation of the electrical efficiency of the FFCTH system with and without CO_2 sequestration compared to the electrical efficiency of the standard sCO_2 turbine system with and without CO_2 sequestration. The standard sCO_2 turbine system was assessed with and without carbon-sequestration to assess the penalty for the baseline case. As expected, the electrical efficiency of the FFCTH system and the standard sCO_2 cycle is higher without sequestration than with sequestration.

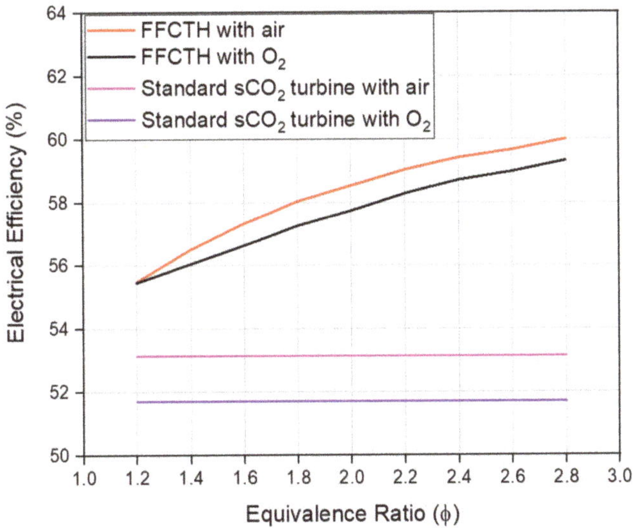

Figure 10. Comparison of the electrical efficiencies of the sCO_2 system with and without the FFC integrated and with and without CO_2 sequestration with varying Φ with experimental FFC parameters ($\eta_{fu} = 0.7$ and $\eta_{fc} = 0.5$).

The electrical efficiency of the FFCTH with and without sequestration changes at different Φ. At lower Φ, the electrical efficiency with and without CO_2 sequestration approach a similar value. This is primarily due to the higher fuel requirement at low Φ (Figure 7) and the power required for CO_2 sequestration in the CO_2 sequestration case compared to no CO_2 sequestration case, as Figure 9 shows. Due to this, the rate of increase of electrical efficiency is much faster for no sequestration case compared to the sequestration case. This shows that the amount of CO_2 generated affects the electrical efficiency reduction in the CO_2 sequestration case, which is in line with logic.

The maximum absolute efficiency penalty is 0.68% at $\Phi = 2.8$. This is a very small efficiency penalty compared to previous literature [36,37,48]. The electrical efficiency of the FFCTH without sequestration is 6.85% higher than the standard sCO_2 turbine setup without sequestration. The standard sCO_2 turbine setup pays a penalty of 1.44% for incorporating CO_2 sequestration.

Table 7 provides a comparison of previous attempts at integrating carbon sequestration with gas turbines. As shown from previous studies, conventional gas turbines typically have a 4–10% decrease in electrical efficiency when integrating post exhaust carbon capture and sequestration. Note that different systems use a variety of CO_2 sequestration methods, which leads to inconsistent efficiency penalties. The standard sCO_2 turbine suffers a smaller penalty in electrical efficiency based on the assumptions in this study, which include the exhaust heated air separation. As the waste heat rejection from the sCO_2 cycle is able to meet the air separation requirements in most cases, the efficiency penalty is small compared to previous work. Along with the fact that FFCTH system pays a much smaller penalty in efficiency; this system also provides an advantage over previous systems due to the high overall efficiency while being sequestration-ready.

Table 7. Comparison of electrical efficiency changes due to addition of carbon sequestration into the system in previous cases and current case.

Type of Power Generation Technique	Base Case Electrical Efficiency	Electrical Efficiency with CO_2 Sequestration	Reference
Gas turbine	53.67	46.75	[36]
Gas turbine	43.88	35.26	[37]
Gas turbine +SOFC	Maximum of 69.5	Maximum of 65.7	[48]
sCO_2 turbine	53.14	51.7	This work
sCO_2 turbine + FFC	Maximum of 60	Maximum of 59.3	This work

We assumed earlier that the heat rejected by the system from the pre-cooler can provide heat for air separation. It is thus important to consider the P/H of the system to evaluate its ability to efficiently convert the incoming fuel energy to electricity compared to heat and make the system sequestration-ready. Since the parameters of the sCO_2 cycle are constant, the heat rejected from the pre-cooler is also constant at 4.81 MW.

Figure 11 shows the P/H of the proposed system with oxygen and with air and a comparison to a standard sCO_2 turbine system. The P/H of the FFCTH is greater than the standard sCO_2 turbine system at all Φ due to larger power generation. The P/H of the FFCTH with oxygen (sequestration-ready) is much higher than with air (not sequestration-ready) at all Φ. The highest P/H for sequestration-ready is 116 times that without at Φ = 2.8. This is primarily because the oxygen separation uses most of the rejected heat from the sCO_2 cycle. Thus, minimum heat remains.

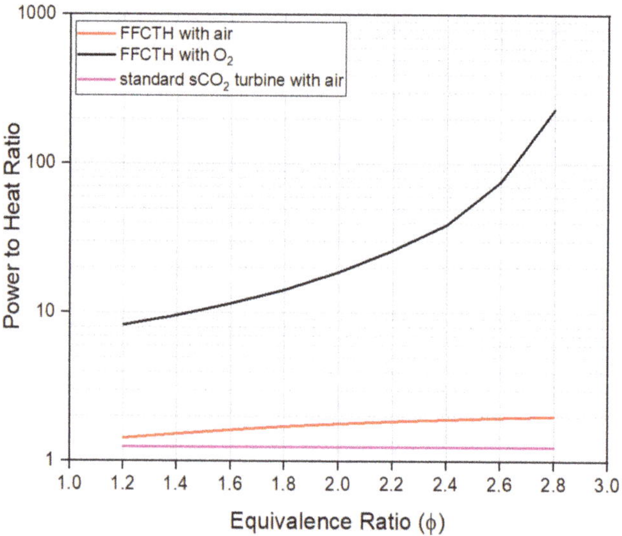

Figure 11. P/H ratios of the system with and without the FFC integrated with and without being CO_2 sequestration-ready at various Φ.

5. Conclusions and Future Work

In this paper, we provided a comprehensive analysis of a novel FFCTH. The system is novel in terms of its attractive efficiency without but also with sequestration-ready CO_2 emissions. The FFC provides heat for the sCO_2 cycle as opposed to a combustor in the standard sCO_2 cycle. Experimental characterization of the FFC assessed power density and efficiency. It also provided FFC parameters, such as fuel utilization efficiency and fuel cell conversion efficiency, for a scaled up model and theoretical analysis.

The electrical efficiency of the FFCTH increases with increasing Φ with and without being CO_2 sequestration-ready. The electrical efficiency comparison of the FFCTH between sequestration-ready and not is only 0.68% lower at the Φ of 2.8. Surprisingly, this electrical efficiency is almost similar with and without CO_2 sequestration readiness at the Φ of 1.2 (0.03% lower with CO_2 sequestration ready case). The close match between the two cases occurs because waste heat can be utilized from the sCO_2 cycle. The electrical efficiency of the FFCTH reaches a maximum electrical efficiency of 60% at a Φ of 2.8. Although a penalty of lower electrical efficiency occurs with CO_2 sequestration, the proposed concept shows a lower reduction in electrical efficiency for carbon sequestration-ready power generation compared to previous literature due to the exhaust heat driven air separation. The proposed system suffers only a 0.68% penalty due to CO_2 sequestration. The results show that the syngas present in the fuel-rich combustion exhaust and the amount of CO_2 present in the exhaust are important factors in describing the results obtained.

The P/H ratio increases with increasing Φ. Furthermore, both with and without being sequestration-ready have greater P/Hs than the standard sCO_2 cycle. The P/H ratio of the FFCTH is 5.7 times higher at the Φ of 1.2 and 116 times higher at the Φ of 2.8 with CO_2 sequestration than without sequestration. The FFCTH minimizes unutilized heat from the system by using the rejected heat from sCO_2 turbine cycle. This shows a wide range of P/H ratios can be achieved by tuning system variables.

Several future studies can build upon this work. As examples, pressure loss, oxygen adsorption/desorption system size and issues with scaling up the experimental results can be addressed. Although the effect of pressure drop is not expected to significantly affect the conclusions of this study, it should be included in future analysis as it will reduce the performance slightly. The experimental results demonstrate the performance of a single fuel cell, but interconnect and other system losses can be considered with a scaled up experiment. In this study the use of waste heat was not considered except for the thermally-driven adsorption/desorption cycle. Other applications include integration with a steam/organic Rankine cycle, or for process heat, both of which can be investigated further.

Author Contributions: Conceptualization, E.B.S., I.E. and R.J.M.; Data curation, R.G.; Formal analysis, R.J.M.; Funding acquisition, E.B.S. and I.E.; Methodology, R.G., E.B.S. and R.J.M.; Project administration, R.J.M.; Supervision, R.J.M.; Writing—original draft, R.G.; Writing—review and editing, R.G., E.B.S., I.E. and R.J.M. All authors have read and agreed to the published version of the manuscript.

Funding: This research was funded by the USA Department of Energy's Office of Energy Efficiency and Renewable Energy (EERE) under the Solar Energy Technologies Office grant number DE-EE0008991.

Acknowledgments: This material is based upon work supported by the USA Department of Energy's Office of Energy Efficiency and Renewable Energy (EERE) under the Solar Energy Technologies Office Award Number DE-EE0008991. The views expressed herein do not necessarily represent the views of the USA Department of Energy or the USA Government.

Conflicts of Interest: The authors declare the following financial interests/personal relationships which may be considered as potential competing interests: A related invention disclosure was submitted.

References

1. U.S. Energy Information Administration. *Monthly Energy Review—January 2020*; 2020; Volume 35. Available online: https://www.eia.gov/totalenergy/data/monthly/pdf/mer.pdf (accessed on 23 September 2020).
2. Lueken, R.; Klima, K.; Griffin, W.M.; Apt, J. The climate and health effects of a USA switch from coal to gas electricity generation. *Energy* **2016**, *109*, 1160–1166. [CrossRef]
3. Bao, C.; Wang, Y.; Feng, D.; Jiang, Z.; Zhang, X. Macroscopic modeling of solid oxide fuel cell (SOFC) and model-based control of SOFC and gas turbine hybrid system. *Prog. Energy Combust. Sci.* **2018**, *66*, 83–140. [CrossRef]
4. Levi, M. Climate consequences of natural gas as a bridge fuel. *Clim. Chang.* **2013**, *118*, 609–623. [CrossRef]

5. De Gouw, J.A.; Parrish, D.D.; Frost, G.J.; Trainer, M. Reduced emissions of CO_2, NOx, and SO_2 from U.S. power plants owing to switch from coal to natural gas with combined cycle technology. *Earth's Future* **2014**, *2*, 75–82. [CrossRef]
6. Hayhoe, K.; Kheshgi, H.S.; Jain, A.K.; Wuebbles, D.J. Substitution of natural gas for coal: Climatic effects of utility sector emissions. *Clim. Chang.* **2002**, *54*, 107–139. [CrossRef]
7. Venkatesh, A.; Jaramillo, P.; Griffin, W.M.; Matthews, H.S. Implications of changing natural gas prices in the United States electricity sector for SO_2, NO_X and life cycle GHG emissions. *Environ. Res. Lett.* **2012**, *7*, 034018. [CrossRef]
8. Bell, M.L.; Dominici, F. Effect modification by community characteristics on the short-term effects of ozone exposure and mortality in 98 US communities. *Am. J. Epidemiol.* **2008**, *167*, 986–997. [CrossRef]
9. Bell, M.L.; Ebisu, K.; Peng, R.D.; Walker, J.; Samet, J.M.; Zeger, S.L.; Dominici, F. Seasonal and Regional Short-term Effects of Fine Particles on Hospital Admissions in 202 US Counties, 1999–2005. *Am. J. Epidemiol.* **2008**, *168*, 1301–1310. [CrossRef]
10. Laden, F.; Schwartz, J.; Speizer, F.E.; Dockery, D.W. Reduction in fine particulate air pollution and mortality: Extended follow-up of the Harvard Six Cities Study. *Am. J. Respir. Crit. Care Med.* **2006**, *173*, 667–672. [CrossRef]
11. Pope, C.A.; Ezzati, M.; Dockery, D.W. Fine-particulate air pollution and life expectancy in the United States. *N. Engl. J. Med.* **2009**, *360*, 376–386. [CrossRef]
12. U.S. Environmental Protection Agency Office of Air and Radiation. *Regulatory Impact Analysis for the Final Clean Air Interstate Rule*; 2005. Available online: https://www.epa.gov/sites/production/files/2020-07/documents/transport_ria_final-clean-air-interstate-rule_2005-03.pdf (accessed on 23 September 2020).
13. Interconnection P. Coal Capacity at Risk for Retirement in PJM: Potential Impacts of the Finalized EPA Cross State Air Pollution Rule and Proposed National Emissions Standards for Hazardous Air Pollutants. 2011. Available online: http://www.psc.ky.gov/PSCSCF/2011%20cases/2011-00401/Kentucky%20Power%20Responses%20to%20042312%20Order/AG/012712/AG%201-14%20Attachments/AG%201-14%20Attachment%207.pdf (accessed on 23 September 2020).
14. Gaudernack, B.; Lynum, S. Natural gas utilisation without CO_2 emissions. *Energy Convers. Manag.* **1997**, *38*, S165–S172. [CrossRef]
15. O'Hayre, R.; Cha, S.-W.; Colella, W.; Prinz, F.B. *Fuel Cell Fundamentals*; Intergovernmental Panel on Climate Change, Ed.; John Wiley & Sons, Inc.: Hoboken, NJ, USA, 2016; ISBN 9781119191766.
16. Milcarek, R.J.; Garrett, M.J.; Welles, T.S.; Ahn, J. Performance investigation of a micro-tubular flame-assisted fuel cell stack with 3,000 rapid thermal cycles. *J. Power Sources* **2018**, *394*, 86–93. [CrossRef]
17. Du, Y.; Finnerty, C.; Jiang, J. Thermal Stability of Portable Microtubular SOFCs and Stacks. *J. Electrochem. Soc.* **2008**, *155*, B972. [CrossRef]
18. Wang, Y.; Zeng, H.; Cao, T.; Shi, Y.; Cai, N.; Ye, X.; Wang, S. Start-up and operation characteristics of a flame fuel cell unit. *Appl. Energy* **2016**, *178*, 415–421. [CrossRef]
19. Wang, Y.; Shi, Y.; Ni, M.; Cai, N. A micro tri-generation system based on direct flame fuel cells for residential applications. *Int. J. Hydrogen Energy* **2014**, *39*, 5996–6005. [CrossRef]
20. Kronemayer, H.; Barzan, D.; Horiuchi, M.; Suganuma, S.; Tokutake, Y.; Schulz, C.; Bessler, W.G. A direct-flame solid oxide fuel cell (DFFC) operated on methane, propane, and butane. *J. Power Sources* **2007**, *166*, 120–126. [CrossRef]
21. Vogler, M.; Horiuchi, M.; Bessler, W.G. Modeling, simulation and optimization of a no-chamber solid oxide fuel cell operated with a flat-flame burner. *J. Power Sources* **2010**, *195*, 7067–7077. [CrossRef]
22. Horiuchi, M.; Suganuma, S.; Watanabe, M. Electrochemical Power Generation Directly from Combustion Flame of Gases, Liquids, and Solids. *J. Electrochem. Soc.* **2004**, *151*, A1402. [CrossRef]
23. Wang, K.; Milcarek, R.J.; Zeng, P.; Ahn, J. Flame-assisted fuel cells running methane. *Int. J. Hydrogen Energy* **2015**, *40*, 4659–4665. [CrossRef]
24. Milcarek, R.J.; Garrett, M.J.; Wang, K.; Ahn, J. Micro-tubular flame-assisted fuel cells running methane. *Int. J. Hydrogen Energy* **2016**, *41*, 20670–20679. [CrossRef]
25. Milcarek, R.J.; Wang, K.; Falkenstein-Smith, R.L.; Ahn, J. Micro-tubular flame-assisted fuel cells for micro-combined heat and power systems. *J. Power Sources* **2016**, *306*, 148–151. [CrossRef]
26. Milcarek, R.J.; Ahn, J. Rich-burn, flame-assisted fuel cell, quick-mix, lean-burn (RFQL) combustor and power generation. *J. Power Sources* **2018**, *381*, 18–25. [CrossRef]

27. Milcarek, R.J.; Garrett, M.J.; Baskaran, A.; Ahn, J. Combustion Characterization and Model Fuel Development for Micro-tubular Flame-assisted Fuel Cells. *J. Vis. Exp.* **2016**, *116*, e54638. [CrossRef] [PubMed]
28. Milcarek, R.J.; Nakamura, H.; Tezuka, T.; Maruta, K.; Ahn, J. Microcombustion for micro-tubular flame-assisted fuel cell power and heat cogeneration. *J. Power Sources* **2019**, *413*, 191–197. [CrossRef]
29. Ghotkar, R.; Milcarek, R.J. POWER2019-1852 Integration of Flame-assisted Fuel Cells with a Gas Turbine running Jet-A as fuel. In Proceedings of the ASME 2019 Power Conference collocated the ASME 2019 Nuclear Forum (POWER2019), Salt Lake City, UT, USA, 15–18 July 2019; pp. 1–9.
30. Milcarek, R.J.; Ahn, J. Micro-tubular flame-assisted fuel cells running methane, propane and butane: On soot, efficiency and power density. *Energy* **2019**, *169*, 776–782. [CrossRef]
31. Milcarek, R.J.; DeBiase, V.P.; Ahn, J. Investigation of startup, performance and cycling of a residential furnace integrated with micro-tubular flame-assisted fuel cells for micro-combined heat and power. *Energy* **2020**, *196*, 117148. [CrossRef]
32. Ghotkar, R.; Milcarek, R.J. Investigation of flame-assisted fuel cells integrated with an auxiliary power unit gas turbine. *Energy* **2020**, *204*, 117979. [CrossRef]
33. Olajire, A.A. A review of mineral carbonation technology in sequestration of CO_2. *J. Pet. Sci. Eng.* **2013**, *109*, 364–392. [CrossRef]
34. Gunning, P.J.; Hills, C.D.; Carey, P.J. Accelerated carbonation treatment of industrial wastes. *Waste Manag.* **2010**, *30*, 1081–1090. [CrossRef]
35. Eloneva, S.; Teir, S.; Salminen, J.; Fogelholm, C.J.; Zevenhoven, R. Fixation of CO_2 by carbonating calcium derived from blast furnace slag. *Energy* **2008**, *33*, 1461–1467. [CrossRef]
36. Cormos, C.C. Integrated assessment of IGCC power generation technology with carbon capture and storage (CCS). *Energy* **2012**, *42*, 434–445. [CrossRef]
37. Tola, V.; Pettinau, A. Power generation plants with carbon capture and storage: A techno-economic comparison between coal combustion and gasification technologies. *Appl. Energy* **2014**, *113*, 1461–1474. [CrossRef]
38. Ermanoski, I.; Stechel, E.B. Thermally-driven adsorption/desorption cycle for oxygen pumping in thermochemical fuel production. *Sol. Energy* **2020**, *198*, 578–585. [CrossRef]
39. Bray, J.M.; Schneider, W.F. Potential energy surfaces for oxygen adsorption, dissociation, and diffusion at the Pt(321) surface. *Langmuir* **2011**, *27*, 8177–8186. [CrossRef]
40. Rochau, G.E.; Pasch, J.J.; Cannon, G.; Carlson, M.; Fleming, D.; Kruizenga, A.; Sharpe, R.; Wilson, M. Supercritical CO_2 Brayton Cycles. 2014. Available online: https://www.osti.gov/servlets/purl/1221819 (accessed on 23 September 2020).
41. Seo, Y.; Huh, C.; Lee, S.; Chang, D. Comparison of CO_2 liquefaction pressures for ship-based carbon capture and storage (CCS) chain. *Int. J. Greenh. Gas Control* **2016**, *52*, 1–12. [CrossRef]
42. Milcarek, R.J.; Garrett, M.J.; Ahn, J. Micro-tubular flame-assisted fuel cells. *J. Fluid Sci. Technol.* **2017**, *12*, JFST0021. [CrossRef]
43. McBride, B.J.; Gordon, S.; McBride, B.J. Computer Program for Calculation of Complex Chemical Equilibrium Compositions and Applications. *NASA Ref. Publ. 1311* **1994**, *184*. Available online: https://web.stanford.edu/~{}cantwell/AA284A_Course_Material/AA284A_Resources/NASA_Glenn_Reports/McBride%20and%20Gordon,%20Computer%20Program%20for%20Calculation%20of%20Complex%20Chemical%20Equilibrium%20NASA%20RP%201311%201996%20Part%20II.pdf (accessed on 23 September 2020).
44. Speight, J.G. *Natural Gas*; Elsevier: Amsterdam, The Netherlands, 2007; ISBN 9781933762142.
45. Milcarek, R.J.; Wang, K.; Garrett, M.J.; Ahn, J. Performance Investigation of Dual Layer Yttria-Stabilized Zirconia–Samaria-Doped Ceria Electrolyte for Intermediate Temperature Solid Oxide Fuel Cells. *J. Electrochem. Energy Convers. Storage* **2016**, *13*, 011002. [CrossRef]
46. Zeng, H.; Gong, S.; Shi, Y.; Wang, Y.; Cai, N. Micro-tubular solid oxide fuel cell stack operated with catalytically enhanced porous media fuel-rich combustor. *Energy* **2019**, *179*, 154–162. [CrossRef]

47. Lemmon, E.W.; Bell, I.H.; Huber, M.L.; McLinden, M.O. *NIST Standard Reference Database 23: Reference Fluid Thermodynamic and Transport Properties-REFPROP*; Version 9.1; National Institute of Standards and Technology: Gaithersburg, MD, USA, 2013; Volume 135.
48. Park, S.K.; Kim, T.S.; Sohn, J.L.; Lee, Y.D. An integrated power generation system combining solid oxide fuel cell and oxy-fuel combustion for high performance and CO_2 capture. *Appl. Energy* **2011**, *88*, 1187–1196. [CrossRef]

Article

Design Analysis of Micro Gas Turbines in Closed Cycles

Krzysztof Kosowski * and Marian Piwowarski *

Faculty of Mechanical Engineering, Gdansk University of Technology, Gabriela Narutowicza Street 11/12, 80-233 Gdansk, Poland

* Correspondence: krzysztof.kosowski@pg.edu.pl (K.K.); marian.piwowarski@pg.edu.pl (M.P.); Tel.: +48-58-347-14-29 (M.P.)

Received: 4 October 2020; Accepted: 2 November 2020; Published: 5 November 2020

Abstract: The problems faced by designers of micro-turbines are connected with a very small volume flow rate of working media which leads to small blade heights and a high rotor speed. In the case of gas turbines this limitation can be overcome by the application of a closed cycle with very low pressure at the compressor inlet (lower than atmospheric pressure). In this way we may apply a micro gas turbine unit of accepted efficiency to work in a similar range of temperatures and the same pressure ratios, but in the range of smaller pressure values and smaller mass flow rate. Thus, we can obtain a gas turbine of a very small output but of the efficiency typical of gas turbines with a much higher power. In this paper, the results of the thermodynamic calculations of the turbine cycles are discussed and the designed gas turbine flow parts are presented. Suggestions of the design solutions of micro gas turbines for different values of power output are proposed. This new approach to gas turbine arrangement makes it possible to build a gas turbine unit of a very small output and a high efficiency. The calculations of cycle and gas turbine design were performed for different cycle parameters and different working media (air, nitrogen, hydrogen, helium, xenon and carbon dioxide).

Keywords: closed cycle gas turbine; different working fluids; thermodynamic analysis; design of turbines; design of compressors

---

1. Introduction

Power plants with gas microturbines are intensively developed nowadays [1], including those operating in polygeneration systems [2,3]. It seems that microturbine technology will be frequently applied in distributed power generation [4,5]. This technology has greatly improved its efficiency thanks to the use of electric generators with rare earth magnets [6–8]. On the other hand, efforts are being made to improve the efficiency, durability and cost reduction of the energy systems based on fuel cells [9–11] or photovoltaic cells [12,13].

The development of technology has made it possible to develop Brayton closed cycles, in which various working media can be used [14]. The closed gas turbine cycle was patented in 1935 by J. Ackeret and C. Keller. The first closed cycle with 2 MW oil-fired gas turbine (AK-36) was built in Switzerland in 1939, followed by another six units built in Switzerland and Germany [15,16]. Gas power plants with a closed cycle and air as a working medium showed a very high reliability at that time. The most representative was the 50 MW helium plant installed in Oberhausen which operated from 1975 to 1987 in Germany [17,18]. Work on the progress of these power plants has been suspended due to the dynamic development of gas turbines in open cycles [19]. Currently, these technologies have become of interest again because of the possibility of their application in nuclear power plants [20,21] or in those power stations that use solar energy [22]. In the systems with nuclear reactors, closed cycles with air, CO_2 [23] or helium [24] are considered for application. Other working media, such as nitrogen or xenon, are also taken into account [25]. Supercritical CO_2 power plants are used which allow to obtain a very good efficiency at low temperatures at the turbine inlet [26,27].

In order to increase the efficiency, closed cycle power plants are modified [28]. The parameters of the working medium are optimized [29] and regenerators [30] and intercoolers [31] can sometimes be applied.

A scheme of a simple closed gas turbine cycle is shown in Figure 1a, [32]. The working medium is compressed in the compressor I and then heated in the high temperature heat exchanger IV. Next, hot gases expand in turbine II, which drives electric generator III. After expansion, the temperature of gases decreases in low temperature heat exchanger V and they return to the compressor inlet. In comparison to the simple open cycle, Figure 1b, the combustion chamber is replaced by an externally fired gas heater IV (source of upper temperature), while the gas cooling heat exchanger V replaces the atmosphere as the source of lower temperature.

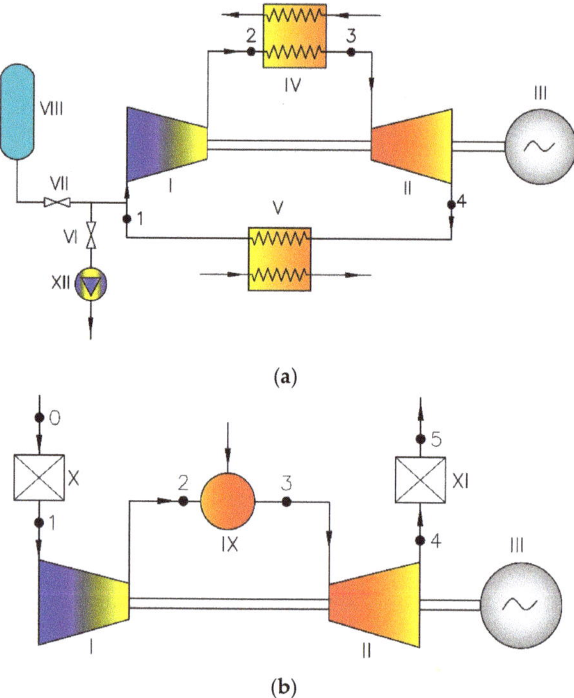

Figure 1. Scheme of a gas turbine simple closed cycle (**a**) and a simple open cycle (**b**); where: I—compressor, II—turbine, III—generator, IV—high temperature exchanger, V—low temperature exchanger, VI and VII—valves, VIII—gas expansion tank, IX—combustion chamber, X—filter, XI—silencer and XII—ejector.

The closed cycle presented in Figure 1a is the simplest possible. More complex cycles contain regeneration, intercooling and reheating systems. The advantages of power plants working in a closed cycle are given below [32]:

- possibility of burning different kinds of fuel (due to an external combustion chamber),
- possibility of using the high pressure of the medium (higher pressure means higher medium density, on the one hand leading to smaller dimensions for the same output, on the other hand enabling the building of high output units),

- possibility of applying different gases as a working medium, in particular gases with better heat exchange parameters and a high value of specific heat c_p, remarkably reducing dimensions and increasing specific turbine output.

The main disadvantage of the closed gas turbine cycle is the necessity of introducing additional heat exchangers, especially the high temperature exchanger IV, which is large in size, heavy and costly (compared with a typical combustion chamber). The low temperature cooler V also increases the dimensions of the whole power plant. Gas turbines can be used in closed cycles with high-temperature gas-cooled reactors (HTGR) and in this form they constitute a competitive alternative to light water nuclear power plants.

As a rule, the main advantage of the closed cycle is the possibility of using the high pressure of the working medium and increasing the output of the power unit. The interpretation of the thermodynamic processes for different power outputs is presented in Figure 2. We propose a reverse situation: a closed cycle with a very low value of pressure at the compressor inlet (lower than the atmospheric pressure). In this way we may apply a gas turbine unit of a relatively high efficiency in the same range of temperatures and the same pressure ratios but in the range of smaller pressure values and a smaller mass flow rate. Thus, we can obtain a gas turbine of a very small output but of the efficiency typical of gas turbines of much higher power.

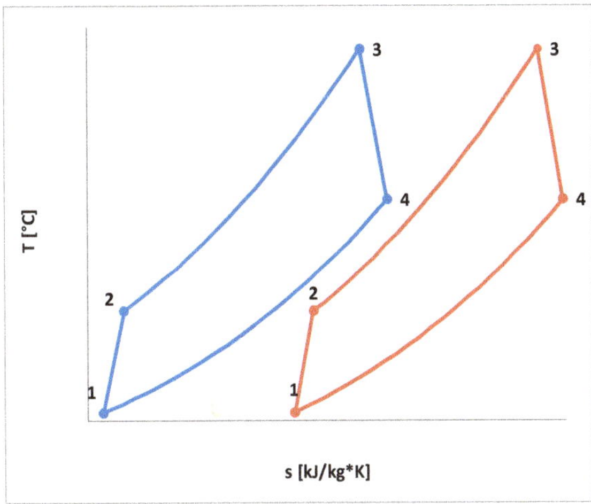

Figure 2. Example of thermodynamic processes interpretation for different power output (gas turbine simple closed cycle).

The most important innovation is that instead of a classical approach to closed gas turbines, a new feature of a closed cycle was shown and applied to design a micro gas turbine of a higher efficiency than the typical engines of the same power. These types of power plants have not been discussed in literature so far.

By applying the proposed method, the micro gas turbines can achieve an efficiency which is higher than that of typical open cycle turbines. For example, the performed analyses have clearly proved that it is possible to design a 10 kW gas micro turbine of a relatively high efficiency of about 31.2% (which is a very competitive value).

2. Modelling

The design analysis was carried out for the turbine closed cycle with a regenerator. The schema of the cycle and its interpretation in a temperature-entropy diagram are shown in Figures 3a and 4, respectively, while the open cycle with a regenerator is presented in Figure 3b.

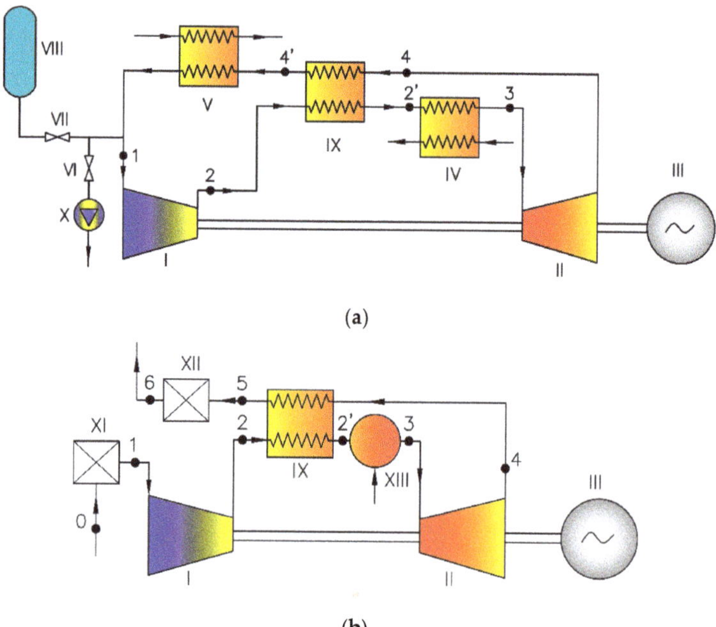

Figure 3. Scheme of a gas turbine closed cycle with regenerator (**a**) and schema of an open cycle with a regenerator (**b**); where: I—compressor, II—turbine, III—generator, IV—high temperature exchanger, V—low temperature exchanger, VI and VII—valves, VIII—gas expansion tank, IX—regenerator, X—ejector, XI—filter, XII—silencer and XIII—combustion chamber.

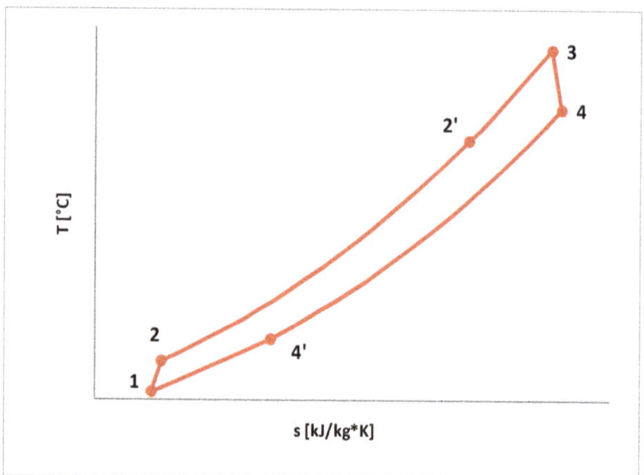

Figure 4. Example of thermodynamic processes interpretation for gas turbine (open or closed) cycle with regenerator.

The classical approach to designing gas turbine plants usually leads to a situation where the decrease in the assumed output results in lower values of volume flow rate, lower diameters, lower blade heights, a drop in the efficiency and an increase in the rotor speed. These relations are illustrated by the comparison of the examples of micro turbines of 50 kW, 35 kW and 10 kW.

The flow parts of the compressors (Figure 5) and the turbines (Figure 6) have been designed by CFD (Computational Fluid Dynamics) codes while the heat exchangers (regenerators) and the combustion chambers have been calculated with the use of standard thermodynamics relations, occasionally applying the iteration method. The symbols and subscripts used in the paper are presented in Tables 1 and 2.

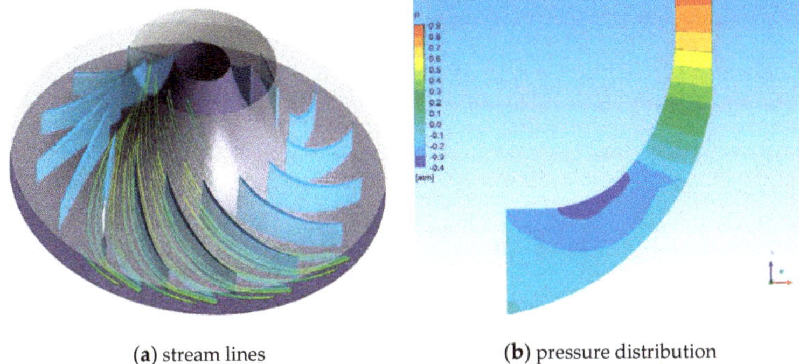

(a) stream lines (b) pressure distribution

Figure 5. Examples of numerical calculations of compressor flow part: stream lines (**a**) and pressure distribution in a channel (**b**).

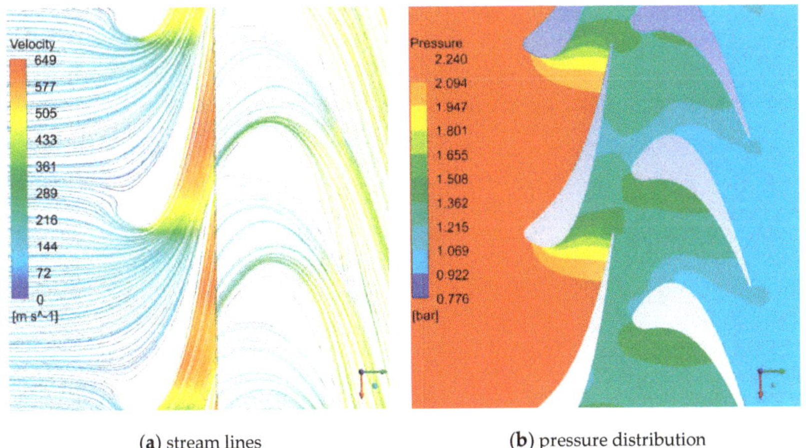

(a) stream lines (b) pressure distribution

Figure 6. Examples of numerical calculations of turbine rotor channels: stream lines (**a**) and pressure distribution (**b**).

Calculations were made with the use of standard formulae (the indices refer to the points marked in the diagram, Figures 1 and 3):

- heat delivered to the working media (heater IV):

$$q_1 = (i_0 - i_6), \tag{1}$$

- heat rejected with the water cooling the heater (V):

$$q_2 = (i_4 - i_1), \tag{2}$$

- turbine work obtained by CFD calculations,
- compressor work obtained by CFD calculations.

Table 1. List of symbols.

Symbol	Description	Unit
D	diameter	[m]
i	enthalpy of unit mass	[kJ/kg]
l	blade height	[m]
m	mass flow rate	[kg/s]
Ma	Mach number	[-]
n	rotational speed	[rpm]
N	power	[kW]
p	pressure	[Pa]
Re	Reynolds number	[-]
Q	heat flux	[kW]
s	entropy of unit mass	[kJ/kg·K]
T	temperature	[°C]
V	volumetric flow rate	[m^3/s]
η	efficiency	[-]
α_1	angle of absolute velocity vector at nozzle exit	[°]
α_2	angle of absolute velocity vector at blade exit	[°]
β_1	angle of relative velocity vector at nozzle exit	[°]
β_2	angle of relative velocity vector at blade exit	[°]
Δ	gradient	[-]
Π	compressor pressure ratio	[-]
σ	regenerator efficiency	[-]

Table 2. List of used subscripts.

Symbol	Description
G	generator
K	compressor
el	electrical
m	mechanical
n	leakage
R	regenerator
T	turbine
D	delivered
$1, 2, \ldots$	numbers on diagrams

The flow parts of the turbine and the compressor have been calculated using ANSYS software. The examples of the distribution of the calculated parameters in flow channels (velocity lines and pressure) are presented in Figures 5 and 6 for the compressor stage and the turbine stage, respectively. The numerical mesh for the compressor flow channel consisted of about 54,000 nodes and 48,600 elements. The example of a mesh at 50% blade span is shown in Figure 7a. Numerical meshes for the turbine nozzles and the rotor blades are shown in Figure 7b,c, respectively. Numerical stationary calculations have been performed assuming the multiple reference method (MRF method), the k-ω shear stress transport (SST) turbulence model and the circumferential symmetry. Air as a working medium has been treated as a viscous and compressible gas.

The energy efficiency is defined as the ratio of the net electric power obtained related to the heat flux supplied to the system and given by the equation:

$$\eta_{el} = \frac{N_{el}}{\dot{Q}_D}. \tag{3}$$

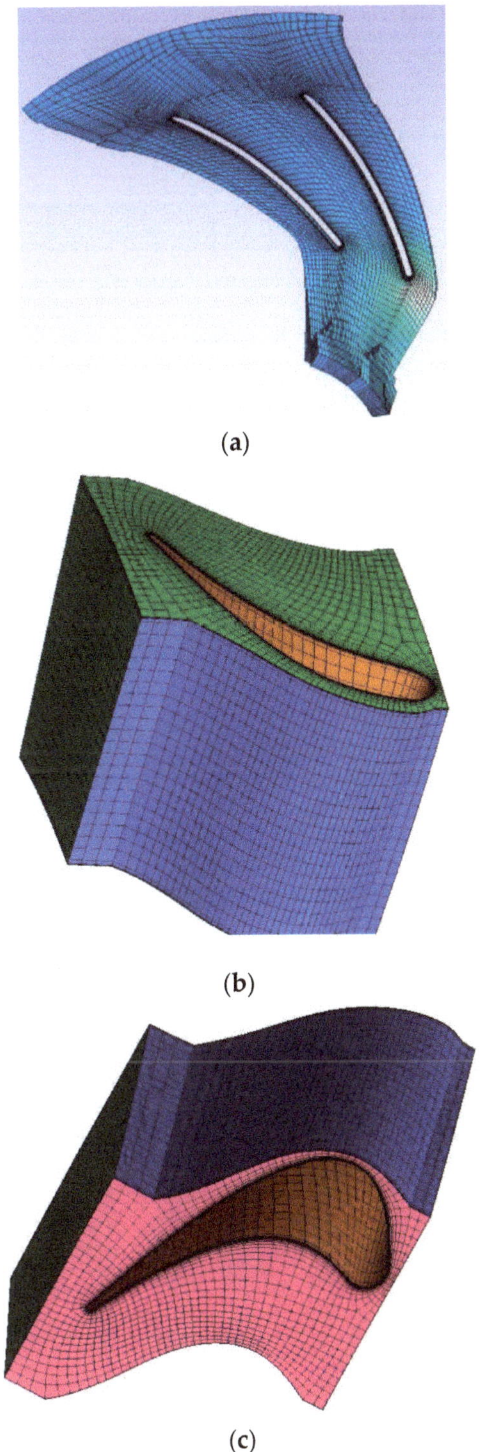

Figure 7. Examples of numerical meshes: compressor channel (**a**), turbine nozzle (**b**) and turbine rotor blade (**c**).

Depending on the variant, the net electric power is the power N_T obtained in the turbine, diminished by the power N_K consumed to drive the compressor and by the losses in the system (generator efficiency η_G, mechanical efficiency η_m, leakage losses).

The effectiveness of the regenerator is defined as the ratio of the actual temperature increase in the exchanger to the maximum possible increase and is determined by the formula:

$$\sigma_R = \frac{\Delta T_{real}}{\Delta T_{\max}} = \frac{T_{2'} - T_2}{T_4 - T_2}. \tag{4}$$

The assumed values of the efficiences or the losses of particular engine elements are presented in Table 3 [32–34]. The chosen numbers can be considered as good average values but not the highest.

Table 3. Assumptions for the design analysis of turbine variants.

Description	Unit	Value
compressor efficiency	[-]	from CFD calculations of particular variant
turbine efficiency	[-]	from CFD calculations of particular variant
mechanical efficiency	[-]	0.980
leakage losses	[-]	0.02
generator efficiency	[-]	0.900
sum of pressure losses	[-]	0.06

The calculations have been performed for temperature T_3 in front of the turbine equal to T_3 = 850 °C (within the range typical of gas turbines of very small output). The parameters of the working media have been determined using REFPROP [35] media library.

3. Results and Discussion

The gas turbine units and the whole power plants of 10 kW, 35 kW and 50 kW were designed. In Figure 8 the example of the turbine power plant is shown, while the main parameters of the particular variants are presented in Table 4 and in Figure 9, the compressor rotors and Figure 10, the turbine discs.

The highest efficiency, equal to about 32%, was obtained for 50 kW turbine, while the lowest (about 23.6%) for 10 kW unit. The 50 kW turbine has the highest diameter, the longest blades and the lowest rotor speed (55,000 rpm), while the 10 kW turbine has the lowest diameter, the smallest blades and the highest rotor speed (120,000 rpm). Thus, the dimensions and the operating parameters vary remarkably for the turbines of different output (Table 4 and Figures 9 and 10). The calculations and the designs have confirmed a well-known statement that the lower the turbine power the more difficult it is to build a turbine of good efficiency.

The situation described above illustrates a typical approach to the designing of gas turbines, however, thanks to the application of a closed cycle with a very low value of pressure at the compressor inlet (lower than the atmospheric pressure), it is possible to obtain a gas turbine of a very small output but with the efficiency typical of gas turbines of much higher power. We may choose a micro gas turbine unit of an accepted design and efficiency to work in the same range of temperatures and the same pressure ratios, but with smaller pressure values and a smaller mass flow rate. Let us consider the gas turbine unit of 50 kW presented above. Applying this engine in a closed cycle and changing the value of pressure at the compressor inlet, it is possible to obtain lower output with the same efficiency. The results are shown in Table 4 and the interpretations of the cycles are presented in Figure 11.

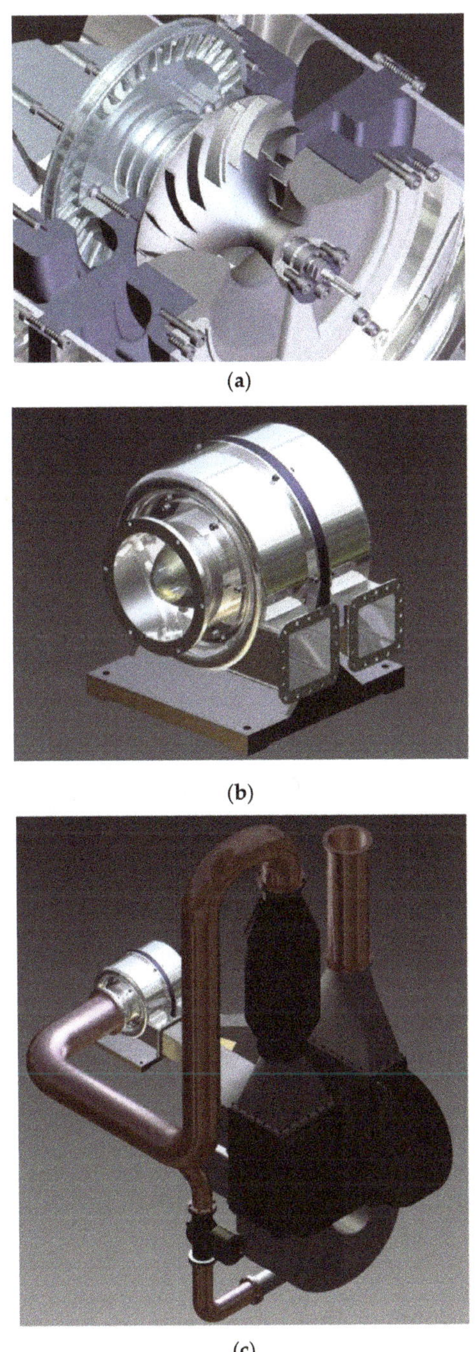

Figure 8. Example of a designed gas turbine power plant: turbine rotor (**a**), turbine unit (**b**) and power plant (**c**).

Table 4. The main parameters of the gas turbine sets.

Description	Unit	Value		
N_e	[kW]	50.0	35.0	10.0
Π	[-]	2.6	2.4	2.2
p_1	[MPa]	0.1000	0.1000	0.1000
T_3	[°C]	850.00	850.00	850.00
η_{REG}	[-]	0.87	0.87	0.88
m_{pow}	[kg/s]	0.5181	0.4722	0.1681
η_{el}	[-]	0.31259	0.26911	0.23665
n	[rpm]	55,000	65,000	120,000
D_{1K}	[m]	0.0790	0.0673	0.0367
D_{2K}	[m]	0.1437	0.1223	0.0667
η_K	[-]	0.83	0.78	0.75
Ma_{w2}	[-]	0.29	0.33	0.32
Ma_{c2}	[-]	0.74	0.73	0.73
Re_{c3}	[-]	326,161.41	327,239.87	335,282.76
α_1	[°]	90.00	90.00	90.00
β_1	[°]	30.00	30.00	30.00
β_2	[°]	44.01	45.12	41.53
α_2	[°]	15.51	15.98	12.92
l_{1K}	[m]	0.0159	0.0139	0.0750
l_{2K}	[m]	0.0085	0.0070	0.0043
D_T	[m]	0.1477	0.1151	0.0688
l_{1T}	[m]	0.0170	0.0164	0.0120
l_{2T}	[m]	0.0180	0.1720	0.0124
η_T	[-]	0.83	0.79	0.76
Ma_{c1}	[-]	0.71	0.73	0.74
Ma_{w2}	[-]	0.74	0.79	0.79
α_1	[°]	17.50	17.00	11.00
β_1	[°]	83.32	70.79	85.11
β_2	[°]	21.78	21.26	14.73
α_2	[°]	87.04	73.96	89.93

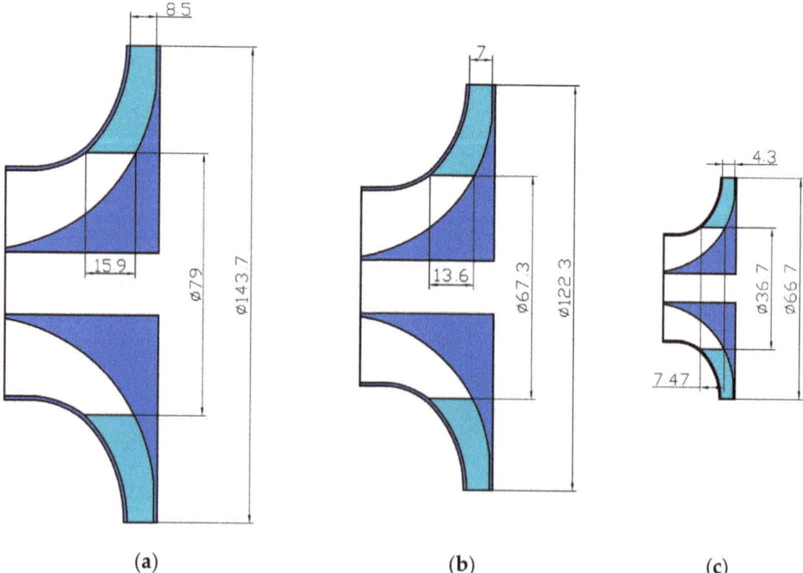

Figure 9. Designed compressor rotors for the gas turbine of 50 kW (**a**), 35 kW (**b**) and 10 kW (**c**).

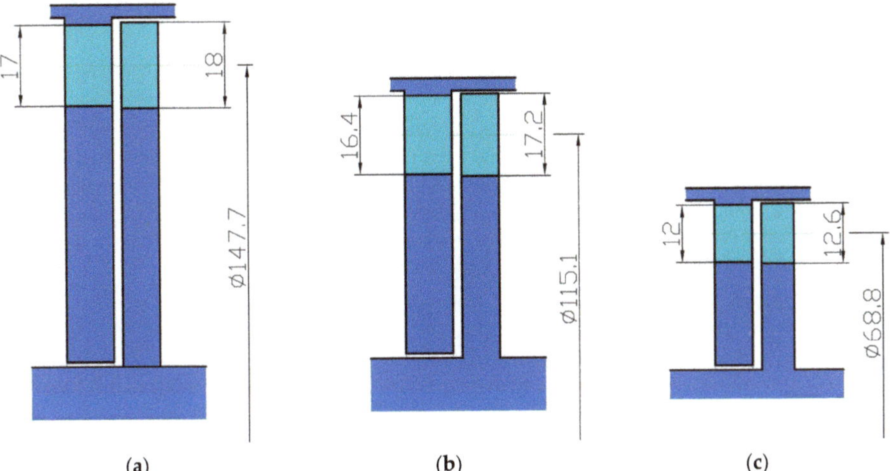

Figure 10. Designed turbine rotors for the gas turbine of 50 kW (**a**), 35 kW (**b**) and 10 kW (**c**).

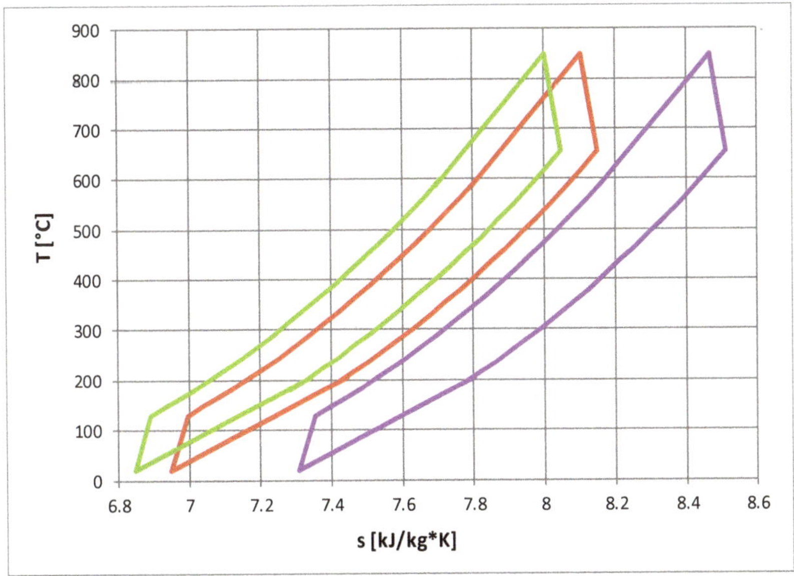

Figure 11. Interpretation of gas turbine closed cycles for different power output, where 50 kW—green line, 35 kW—red line and 10 kW—purple line.

As a result of the calculations, we may conclude that it is enough to apply a highly efficient gas turbine in a closed cycle and reduce the pressure at the compressor inlet in order to obtain a power plant of a smaller output but the same high efficiency. In the presented example of the 50 kW turbine with a reduction of pressure in front of the compressor from 0.1 MPa to 0.07 MPa or to 0.02 MPa allows us to receive 35 kW and 10 kW, respectively. All the variants have the same efficiency (31.26%), identical rotor speed (55,000 rpm), the same pressure ratio (2.6) and the same upper temperature (T_3 = 850 °C) (Table 4). It seems that the proposed method enables us to obtain a gas turbine of a very small output but high efficiency and a relatively small rotor speed. Thus, a new feature of a closed gas turbine cycle was shown and applied to design a micro turbine of a relatively high efficiency. This conclusion is one of the most important innovations that has not been described in literature so far.

Closed gas turbine cycles make it possible to apply different working media, especially the gases with better heat exchange parameters and a high value of specific heat c_p, thus remarkably reducing the dimensions and, at the same time, increasing specific turbine output. The calculations of the closed cycles have been performed for six different fluids: air, nitrogen, helium, carbon dioxide, hydrogen and xenon and for various values of the temperature in front of the turbine (800–1000 °C, the range typical of gas turbines of very small output). The parameters of the working media have been determined using REFPROP [35] media library. In Figure 12 the overall efficiency of the cycles with a regenerator and the upper temperature T_3 = 850 °C are presented as the function of the compressor pressure ratio. The engine efficiency reaches over 36% for air, nitrogen and hydrogen. It is slightly smaller for helium and xenon and about 2% lower for carbon dioxide. The optimum pressure ratio varies from less than 2 for helium and xenon to about 3.7 for carbon dioxide.

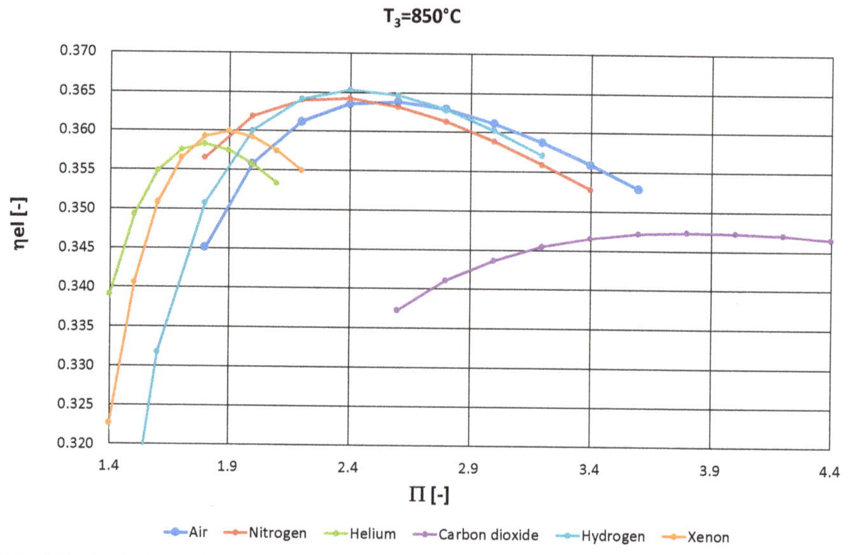

Figure 12. Efficiency of the cycles with regenerator as the function of the compressor pressure ratio.

4. Conclusions

The authors have presented the results of the design analysis of micro turbines proving that it is possible to increase the efficiency of gas turbines of small output. By the application of a closed cycle with a very low value of pressure at the compressor inlet a gas turbine of a very small output is obtained which has the efficiency typical of gas turbines of a much higher power. We may choose a well-designed micro gas turbine unit of high efficiency and apply it to work in the same range of temperatures and the same pressure ratios but with smaller pressure values and a smaller mass flow rate. Thus, a gas turbine of a very small output can operate with the efficiency typical of gas turbines of a much higher power. In the presented example the efficiency of the 10 kW turbine has increased from 23.6% to 31.2%.

The performed analyses have clearly proved that applying closed gas turbine cycles is possible when we use different working media. The cycle and the gas turbine design calculations were performed for different cycle parameters and different working media (air, nitrogen, hydrogen, helium, xenon and carbon dioxide). Their better heat exchange parameters and a higher value of specific heat c_p can reduce the engine dimensions and increase the specific turbine output.

Author Contributions: Conceptualization, K.K. and M.P.; methodology, K.K. and M.P.; formal analysis, K.K. and M.P.; data curation, K.K. and M.P.; writing—original draft preparation, K.K. and M.P.; writing—review and editing, K.K. and M.P.; supervision, K.K. and M.P.; funding acquisition, K.K. and M.P. All authors have read and agreed to the published version of the manuscript.

Funding: This research received financial support from The National Centre for Research and Development (Poland), project: Development of integrated technologies in the production of fuel and energy from biomass, agricultural and other waste.

Acknowledgments: The authors would like to thank Robert Stępień and Wojciech Włodarski from Gdansk University of Technology for their help in performing the ANSYS calculations and preparing the figures.

Conflicts of Interest: The authors declare no conflict of interest. The funders had no role in the design of the study; in the collection, analyses or interpretation of data; in the writing of the manuscript; and in the decision to publish the results.

References

1. Kosowski, K.; Włodarski, W.; Piwowarski, M.; Stępień, R. Performance characteristics of a micro-turbine. *Adv. Vib. Eng.* **2014**, *2*, 341–350.
2. Stępniak, D.; Piwowarski, M. Analyzing selection of low-temperature medium for cogeneration micro power plant. *Pol. J. Environ. Stud.* **2014**, *23*, 1417–1421.
3. Thu, K.; Saha, B.B.; Chua, K.; Bui, T.D. Thermodynamic analysis on the part-load performance of a microturbine system for micro/mini-CHP applications. *Appl. Energy* **2016**, *178*, 600–608. [CrossRef]
4. Kosowski, K.; Stępień, R.; Włodarski, W.; Piwowarski, M.; Hirt, Ł. Partial admission stages of high efficiency for a microturbine. *J. Vib. Eng. Technol.* **2014**, *2*, 441–448.
5. Soares, C. *Microturbines: Applications for Distributed Energy Systems*, 1st ed.; Elsevier's Science & Technology: Oxford, UK, 2007; ISBN 978-0750684699.
6. Włodarski, W. Experimental investigations and simulations of the microturbine unit with permanent magnet generator. *Energy* **2018**, *158*, 59–71. [CrossRef]
7. Włodarski, W. Control of a vapour microturbine set in cogeneration applications. *ISA Trans.* **2019**, *94*, 276–293. [CrossRef]
8. Włodarski, W. A model development and experimental verification for a vapour microturbine with a permanent magnet synchronous generator. *Appl. Energy* **2019**, *252*, 113430. [CrossRef]
9. Saadabadi, S.A.; Thattai, A.T.; Fan, L.; Lindeboom, R.E.; Spanjers, H.; Aravind, P. Solid Oxide Fuel Cells fuelled with biogas: Potential and constraints. *Renew. Energy* **2019**, *134*, 194–214. [CrossRef]
10. Kwaśniewski, T.; Piwowarski, M. Design analysis of hybrid gas turbine-fuel cell power plant in stationary and marine applications. *Pol. Marit. Res.* **2020**, *2*, 107–119. [CrossRef]
11. Cozzolino, R.; Lombardi, L.; Tribioli, L. Use of biogas from biowaste in a solid oxide fuel cell stack: Application to an off-grid power plant. *Renew. Energy* **2017**, *111*, 781–791. [CrossRef]
12. Herrando, M.; Pantaleo, A.M.; Wang, K.; Markides, C.N. Solar combined cooling, heating and power systems based on hybrid PVT, PV or solar-thermal collectors for building applications. *Renew. Energy* **2019**, *143*, 637–647. [CrossRef]
13. Hsu, P.-C.; Huang, B.J.; Wu, P.-H.; Wu, W.-H.; Lee, M.-J.; Yeh, J.-F.; Wang, Y.-H.; Tsai, J.-H.; Li, K.; Lee, K.-Y. Long-term Energy Generation Efficiency of Solar PV System for Self-consumption. *Energy Procedia* **2017**, *141*, 91–95. [CrossRef]
14. Kunniyoor, V.; Singh, P.; Nadella, K. Value of closed-cycle gas turbines with design assessment. *Appl. Energy* **2020**, *269*, 114950. [CrossRef]
15. Bammert, K.; Rurik, J.; Griepentrog, H. Highlights and Future Development of Closed-Cycle Gas Turbines. *J. Eng. Power* **1974**, *96*, 342–348. [CrossRef]
16. Kehlhofer, R.; Kunze, N.; Lehmann, J.; Schüller, K.H. *Gasturbinenkraftwerke, Kombikraftwerke, Heizkraftwerke und Industriekraftwerke Resch, Handbuchreihe Energie*; Band 7; Technischer Verlag Resch: München, Germany, 1992; ISBN 3-88585-094-X.
17. McDonald, C.F.; Boland, C.R. The Nuclear Closed-Cycle Gas Turbine (HTGR-GT)—Dry Cooled Commercial Power Plant Studies. *J. Eng. Power* **1981**, *103*, 89–100. [CrossRef]
18. Wolf, J.; Perkavec, M. *Neuste Entwicklungen im Gasturbinenbau—Gasturbinen in Energetischen Anlagen*; VDI Berichte: Leverkusen, Deutchland, 1998.

19. Ahn, Y.; Lee, J.I. Study of various Brayton cycle designs for small modular sodium-cooled fast reactor. *Nucl. Eng. Des.* **2014**, *276*, 128–141. [CrossRef]
20. Bae, S.J.; Lee, J.; Ahn, Y.; Lee, J.I. Preliminary studies of compact Brayton cycle performance for Small Modular High Temperature Gas-cooled Reactor system. *Ann. Nucl. Energy* **2015**, *75*, 11–19. [CrossRef]
21. Malik, A.; Zheng, Q.; Qureshi, S.R.; Zaidi, A.A. Effect of helium xenon as working fluid on the compressor of power conversion unit of closed Brayton cycle HTGR power plant. *Int. J. Hydrogen Energy* **2020**, *45*, 10119–10129. [CrossRef]
22. Stein, W.H.; Buck, R. Advanced power cycles for concentrated solar power. *Sol. Energy* **2017**, *152*, 91–105. [CrossRef]
23. Kizilkan, O.; Khanmohammadi, S.; Saadat-Targhi, M. Solar based CO2 power cycle employing thermoelectric generator and absorption refrigeration: Thermodynamic assessment and multi-objective optimization. *Energy Convers. Manag.* **2019**, *200*, 112072. [CrossRef]
24. McDonald, C.F. Helium turbomachinery operating experience from gas turbine power plants and test facilities. *Appl. Therm. Eng.* **2012**, *44*, 108–142. [CrossRef]
25. Tournier, J.-M.P.; El-Genk, M.S. Properties of noble gases and binary mixtures for closed Brayton Cycle applications. *Energy Convers. Manag.* **2008**, *49*, 469–492. [CrossRef]
26. Liese, E.; Albright, J.; Zitney, S.A. Startup, shutdown, and load-following simulations of a 10 MWe supercritical CO2 recompression closed Brayton cycle. *Appl. Energy* **2020**, *277*, 115628. [CrossRef]
27. Ma, Y.; Liu, M.; Yan, J.; Liu, J. Performance investigation of a novel closed Brayton cycle using supercritical CO2-based mixture as working fluid integrated with a LiBr absorption chiller. *Appl. Therm. Eng.* **2018**, *141*, 531–547. [CrossRef]
28. Olumayegun, O.; Wang, M.; Kelsall, G. Closed-cycle gas turbine for power generation: A state-of-the-art review. *Fuel* **2016**, *180*, 694–717. [CrossRef]
29. Liu, H.; Chi, Z.; Zang, S. Optimization of a closed Brayton cycle for space power systems. *Appl. Therm. Eng.* **2020**, *179*, 115611. [CrossRef]
30. Sadatsakkak, S.A.; Ahmadi, M.H.; Ahmadi, M.A. Thermodynamic and thermo-economic analysis and optimization of an irreversible regenerative closed Brayton cycle. *Energy Convers. Manag.* **2015**, *94*, 124–129. [CrossRef]
31. Miao, H.; Wang, Z.; Niu, Y. Key issues and cooling performance comparison of different closed Brayton cycle based cooling systems for scramjet. *Appl. Therm. Eng.* **2020**, *179*, 115751. [CrossRef]
32. Kosowski, K. *Steam and Gas Turbines with the Examples of Alstom Technology*; Alstom: Saint-Ouen, France, 2007; ISBN 978-83-925959-3-9.
33. Mikielewicz, D.; Kosowski, K.; Tucki, K.; Piwowarski, M.; Stępień, R.; Orynycz, O.; Włodarski, W. Gas Turbine Cycle with External Combustion Chamber for Prosumer and Distributed Energy Systems. *Energies* **2019**, *12*, 3501. [CrossRef]
34. Mikielewicz, D.; Kosowski, K.; Tucki, K.; Piwowarski, M.; Stępień, R.; Orynycz, O.; Włodarski, W. Influence of Different Biofuels on the Efficiency of Gas Turbine Cycles for Prosumer and Distributed Energy Power Plants. *Energies* **2019**, *12*, 3173. [CrossRef]
35. Lemmon, E.; Huber, M.; McLinden, M. NIST Standard Reference Database 23: Reference Fluid Thermodynamic and Transport Properties-REFPROP. In *Standard Reference Data Program*; Version 9.0 edition; National Institute of Standards and Technology: Gaithersburg, MD, USA, 2010.

MDPI
St. Alban-Anlage 66
4052 Basel
Switzerland
Tel. +41 61 683 77 34
Fax +41 61 302 89 18
www.mdpi.com

Energies Editorial Office
E-mail: energies@mdpi.com
www.mdpi.com/journal/energies

www.ingramcontent.com/pod-product-compliance
Lightning Source LLC
LaVergne TN
LVHW070642170726
843515LV00004B/302

* 9 7 8 3 0 3 6 5 0 0 7 2 0 *